Studies in Computational Intelligence

Volume 1091

Series Editor

Janusz Kacprzyk, Polish Academy of Sciences, Warsaw, Poland

The series "Studies in Computational Intelligence" (SCI) publishes new developments and advances in the various areas of computational intelligence—quickly and with a high quality. The intent is to cover the theory, applications, and design methods of computational intelligence, as embedded in the fields of engineering, computer science, physics and life sciences, as well as the methodologies behind them. The series contains monographs, lecture notes and edited volumes in computational intelligence spanning the areas of neural networks, connectionist systems, genetic algorithms, evolutionary computation, artificial intelligence, cellular automata, self-organizing systems, soft computing, fuzzy systems, and hybrid intelligent systems. Of particular value to both the contributors and the readership are the short publication timeframe and the world-wide distribution, which enable both wide and rapid dissemination of research output.

Indexed by SCOPUS, DBLP, WTI Frankfurt eG, zbMATH, SCImago.

All books published in the series are submitted for consideration in Web of Science.

Bernd-Holger Schlingloff · Thomas Vogel ·
Andrzej Skowron
Editors

Concurrency, Specification and Programming

Revised Selected Papers from the 29th
International Workshop on Concurrency,
Specification and Programming (CS&P'21),
Berlin, Germany

 Springer

Editors
Bernd-Holger Schlingloff ⓘ
Institut für Informatik
Humboldt Universität zu Berlin
Berlin, Germany

Thomas Vogel ⓘ
Department of Computer Science
Humboldt-Universität zu Berlin
Berlin, Germany

Andrzej Skowron ⓘ
Systems Research Institute
Polish Academy of Sciences
Warsaw, Poland

ISSN 1860-949X ISSN 1860-9503 (electronic)
Studies in Computational Intelligence
ISBN 978-3-031-26653-9 ISBN 978-3-031-26651-5 (eBook)
https://doi.org/10.1007/978-3-031-26651-5

This Springer imprint is published by the registered company Springer Nature Switzerland AG
The registered company address is: Gewerbestrasse 11, 6330 Cham, Switzerland

Preface

The CS&P workshop series deals with the formal specification of concurrent and parallel systems, mathematical models for describing such systems and programming and verification concepts for their implementation. It is an annual seminar formerly organized every even year by Humboldt Universität zu Berlin and every odd year by Warsaw University. Due to the corona pandemic, this order has now been inverted. The workshop has a tradition dating back to the mid-seventies; since 1993 it was named CS&P. During almost 30 years, CS&P has become an important forum for researchers, especially from European and Asian countries.

In 2020, the tradition was interrupted; there was no CS&P workshop, and CS&P'20 was postponed to September 2021 and renamed to CS&P'21: the 29th International Workshop on Concurrency, Specification and Programming 2021. The event was held hybrid, both as an online event and as an on-site conference with physical meeting of the participants. The on-site meeting took place at GFaI e.V., the society for the advancement of applied informatics (Gesellschaft zur Förderung der angewandten Informatik), in Berlin-Adlershof, Germany.

The program of CS&P'21 comprised two invited keynotes and presentations of 16 peer-reviewed papers. All talks reflect the current trends in the CS&P field: there are "classical" contributions on the theory of concurrency, specification and programming such as event/data-based systems, cause-effect structures, granular computing, and time Petri nets. However, more and more papers are also concerned with artificial intelligence and machine learning techniques, e.g., in the set of pairs of objects, or with sparse neural networks. Moreover, application areas such as the prediction of football game results, the classification of dry beans or the care for honeybees are the focus of attention. Furthermore, this volume contains several papers on software quality assurance, e.g., fault localization and automated testing of software-based systems.

After CS&P'21, we invited the authors of 12 papers and the keynote speakers to contribute to this book by revising and extending their papers. We received ten submissions, each of which has been peer-reviewed in a second round by program committee members of CS&P'21. Finally, nine papers have been accepted for publication in this book. This book provides a high-quality collection of papers on concurrency, specification and programming.

We further thank Prof. Janusz Kacprzyk, Dr. Thomas Ditzinger and the staff of the Series Computational Intelligence Springer for their support in making this book possible.

Altogether, we hope that you will find the collection to constitute an interesting read and offer many connecting factors for further research.

Berlin, Warsaw   Bernd-Holger Schlingloff
December 2022   Andrzej Skowron
                                                                                 Thomas Vogel

Acknowledgements

We would like to thank the authors of the articles included in the book and all the reviewers Ludwik Czaja, Lars Grunske, Andrzej Janusz, Magdalena Kacprzak, Lech Polkowski, Edip Senyurek, Marcin Szczuka, Matthias Weidlich, Matthias Werner and Dmitry Zaitsev whose work contributed significantly to improving the quality of the book.

We extend an expression of gratitude to Prof. Janusz Kacprzyk, to Dr. Thomas Ditzinger and to the Series Computational Intelligence staff at Springer for their support in making this book possible.

Contents

Contributors

Agnieszka Boruta Warsaw University of Live Sciences, Warsaw, Poland. e-mail: agnieszka_boruta@sggw.edu.pl

Elena Bozhenkova A.P. Ershov Institute of Informatics Systems, Novosibirsk, Russia. e-mail: bozhenko@iis.nsk.su

Soma Dutta University of Warmia and Mazury in Olsztyn, Olsztyn, Poland. e-mail: somadutta9@gmail.com

Pawel Gburzynski Vistula University, Warsaw, Poland. e-mail: p.gburzynski@vistula.edu.pl

Paweł Gora Institute of Informatics, Faculty of Mathematics, Informatics and Mechanics, University of Warsaw, Warsaw, Poland. e-mail: p.gora@mimuw.edu.pl

Robin Gröpler ifak–Institut für Automation und Kommunikation e.V., Magdeburg, Germany. e-mail: robin.groepler@ifak.eu

Damas P. Gruska Department of Applied Informatics, Comenius University, Bratislava, Slovak Republic. e-mail: gruska@fmph.uniba.sk

Marek Grzegorowski Institute of Informatics, University of Warsaw, Warsaw, Poland. e-mail: m.grzegorowski@mimuw.edu.pl

Verena V. Hafner Humboldt-Universität zu Berlin, Unter den Linden 6, Berlin, Germany. e-mail: hafner@informatik.hu-berlin.de

Eyad Kannout Institute of Informatics, University of Warsaw, Warsaw, Poland. e-mail: eyad.kannout@mimuw.edu.pl

Arkadiusz Klemenko Garbary 95B/95, Poznań, Poland. e-mail: arek.klemenko@gmail.com

Libin Kutty ifak–Institut für Automation und Kommunikation e.V., Magdeburg, Germany. e-mail: libinjohn2@gmail.com

Ewa Kuznicka Warsaw University of Live Sciences, Warsaw, Poland. e-mail: ewa_kuznicka@sggw.edu.pl

Heinrich Mellmann Humboldt-Universität zu Berlin, Unter den Linden 6, Berlin, Germany. e-mail: mellmann@informatik.hu-berlin.de

Marcin Możejko Institute of Informatics, Faculty of Mathematics, Informatics and Mechanics, University of Warsaw, Warsaw, Poland. e-mail: mmozejko1988@gmail.com

Roman R. Redziejowski Ceremonimästarvägen 10, Lidingö, Sweden. e-mail: roman@redz.se

Łukasz Skowronek Brygadzistów 23a/3, Katowice, Poland. e-mail: lukasz.m.skowronek@gmail.com

Hung Son Nguyen Institute of Informatics, University of Warsaw, Warsaw, Poland. e-mail: son@mimuw.edu.pl

Viju Sudhi ifak–Institut für Automation und Kommunikation e.V., Magdeburg, Germany. e-mail: vjusudhi@gmail.com

Volha Taliaronak Humboldt-Universität zu Berlin, Unter den Linden 6, Berlin, Germany. e-mail: taliarov@informatik.hu-berlin.de

Irina Virbitskaite A.P. Ershov Institute of Informatics Systems, Novosibirsk, Russia. e-mail: virb@iis.nsk.su

Chapter 1
Natural Language Processing for Requirements Formalization: How to Derive New Approaches?

Viju Sudhi, Libin Kutty, and Robin Gröpler

Abstract It is a long-standing desire of industry and research to automate the software development and testing process as much as possible. Model-based design and testing methods have been developed to automate several process steps and handle the growing complexity and variability of software systems. However, major effort is still required to create specification models from a large set of functional requirements provided in natural language. Numerous approaches based on natural language processing (NLP) have been proposed in the literature to generate requirements models using mainly syntactic properties. Recent advances in NLP show that semantic quantities can also be identified and used to provide better assistance in the requirements formalization process. In this work, we present and discuss principal ideas and state-of-the-art methodologies from the field of NLP in order to guide the readers on how to derive new requirements formalization approaches according to their specific use case and needs. We demonstrate our approaches on two industrial use cases from the automotive and railway domains and show that the use of current pre-trained NLP models requires less effort to adapt to a specific use case. Furthermore, we outline findings and shortcomings of this research area and propose some promising future developments.

1.1 Introduction

Requirements engineering (RE) plays a fundamental role for the software development and testing process. There are usually many people involved in this process, such as the customer or sponsor, the users from different areas of expertise, the development and testing team, and those responsible for the system architecture. Therefore, requirements are intentionally written in a textual form in order to be understandable for all stakeholders of the software product to be developed. However, this also means that requirements have an inherently informal character due to the ambiguity

V. Sudhi · L. Kutty · R. Gröpler (✉)
ifak–Institut für Automation und Kommunikation e.V., Magdeburg, Germany
e-mail: robin.groepler@ifak.eu

© The Author(s), under exclusive license to Springer Nature Switzerland AG 2023
B. Schlingloff et al. (eds.), *Concurrency, Specification and Programming, Studies in Computational Intelligence 1091*, https://doi.org/10.1007/978-3-031-26651-5_1

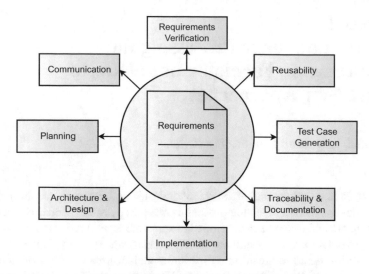

Fig. 1.1 Requirements as the basis for the software development and testing process

and diversity of human language. This makes it difficult to automatically analyze the requirements for further processing. There are many different processes that build on them, such as requirements verification and test case generation (see Fig. 1.1).

In order to handle the growing complexity and variability of software systems, model-based design and testing methods have been developed [3]. Especially for safety-critical systems such as in the automotive, railway and aerospace domains, extensive testing based on the requirements is necessary. However, the manual creation of requirements models from natural language requirements is time-consuming and error-prone, and also requires a lot of expert knowledge. This is especially true in agile software development which involves continuous improvements and many changes to requirements.

Numerous approaches based on natural language processing (NLP) have been proposed in the literature to generate requirements models using mainly syntactic properties [20, 40]. Recent advances in NLP show that semantic quantities can also be identified and used to provide better assistance in the formalization of unrestricted natural language requirements [13, 28]. However, most studies propose concrete solutions that work well only for a very specific environment.

The aim of this work is to focus on principal ideas and state-of-the-art methodologies from the field of NLP to automatically generate requirements models from natural language requirements. We iteratively derive a set of rules based on NLP information in order to guide readers on how to create their own set of rules and methods according to their specific use cases and needs. In particular, we want to investigate the question: *How to derive new approaches for requirements formalization using natural language processing?* We highlight and discuss the necessary stages of an NLP based pipeline towards the generation of requirements models. We present and discuss two approaches in detail: (i) *a dependency- and part-of-speech-based approach*, and (ii) *a semantic-role-based approach*. This significantly extends and enhances our previous work [15]. The approaches are demonstrated on two industrial use cases: a battery charging approval system from the automotive domain and a propulsion control system in the railway domain. The requirements models are represented in a human- and machine-readable format in a form of pseudocode [16]. In summary, this work aims to provide more general and long-lasting instructions on how to develop new approaches for NLP-based requirements formalization.

In Sect. 1.2, we present our proposed NLP-based pipeline, introduce the use cases, and present the two different approaches for NLP-based requirements formalization. In Sect. 1.3, we review the state of the literature on the various NLP methods for requirements formalization. In Sect. 1.4, we discuss some general findings and shortcomings in this area of research and provide an outlook on possible future developments. Finally, Sect. 1.5 concludes our work.

1.2   Methodology

In this section, we propose an NLP-based pipeline for requirements formalization. We first give a brief overview of the use cases and then present and discuss the two different approaches in detail and demonstrate the iterative development of rule sets.

The automatic generation of requirements models from functional requirements written in unrestricted natural language is highly complex and needs to be divided into several steps. Therefore, our pipeline consists of several stages, as illustrated in Fig. 1.2. The different stages of the pipeline are described in more detail below.

Stage 1: Preprocessing. We start with preprocessing the requirements. According to our use cases, we perform data preparation, where we clean up the raw text, and resolve pronouns. This stage need not be limited to these steps and should be flexibly adapted to the style and domain of the requirements at hand.

Stage 2: Decomposition. In order to extract individual actions, the preprocessed requirements are decomposed into clauses. Industrial requirements tend to be complex in certain cases with multiple sentences combined together to convey a particular system behavior. We decompose each requirement sentence at certain conjunctions and linking words (*if*, *while*, *until*, *and*, *or*, etc.), assuming that each clause contains a single action. Multi-line and multi-paragraph requirements should also be decomposed into clauses.

Stage 3: Entity detection. In this stage, we use the syntactic and semantic information of each clause to identify the desired model entities, such as *signals*, *components*, and *parameters*. We construct a rule-based mapping that is iteratively derived from the considered requirements of our specific use cases. Further, we map comparison words to *operator* symbols ($<$, $>$, $==$, etc.) using a dictionary. These rules can be

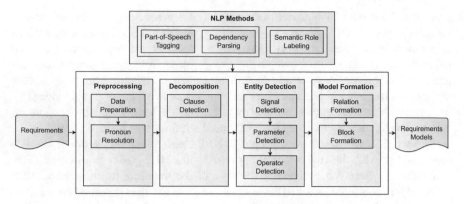

Fig. 1.2 Proposed pipeline for automatic requirements formalization based on NLP information

easily adapted to different use cases and domains by following the same procedure as described below.

Stage 4: Model formation. In the final stage of the pipeline, we assemble the retrieved information to form *relations* for each clause. These can either be assignments of the form `signal(parameter*)` with optional parameter, or conditional statements of the form `signal() operator parameter`. Then we combine them according to the derived logical operators, and assign them to specific *blocks* (*if*, *then*, *else*, etc.). This yields a requirements model for each requirement.

The whole pipeline can be tailored according to the use case and the desired output. For example, no components (actors) are explicitly mentioned in the requirements we consider. Therefore, we have omitted this entity and assume that this information is given. The decomposition and entity detection stages are particularly based on information provided by NLP methods. We present two approaches to handle these stages: (i) *a dependency- and part-of-speech-based approach* (Sect. 1.2.2), which utilizes the grammatical constructs of the requirement text, and (ii) *a semantic-role-based approach* (Sect. 1.2.3), which makes use of the semantic roles given by a pre-trained model. We provide a detailed implementation of both approaches, presented below, along with sample data on GitHub.[1] We emphasize that all stages of the requirements formalization process are intended to be fully automated. We assume that a requirements engineer reviews the generated models and adjust them as necessary. Both approaches work with unrestricted natural language and can be further refined and adapted to different styles and domains according to the needs of the use case.

---

[1] https://github.com/ifak-prototypes/nlp_reform.

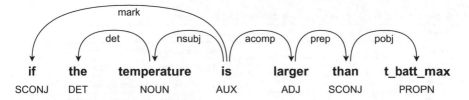

Fig. 1.3 Dependency and POS tags for an exemplary requirement phrase

1.2.1 Use Cases

We demonstrate the derivation of our approaches using functional requirements from two different industrial use cases. The first use case is a battery charging approval system provided by AKKA from the automotive domain. The use case describes a system for charging approval of an electric vehicle in interaction with a charging station. In total, AKKA has provided 14 separate requirement statements. The requirements are used for a model-based software development and implementation of the functionality in an electronic control unit (ECU). More details about the industrial use case can be found in [15, 17].

The second industrial use case is a propulsion control system (PPC) provided by Alstom from the railway domain. The PPC is part of a large, complex, safety-critical system. It handles the control of the entire propulsion system, including both control software as well as the electrical functions. Alstom has provided 31 requirements which do not follow a prescribed format in order not to focus on syntax when writing them. The system architecture and software modules are modeled in Matlab Simulink using a model-based design approach. More information about this use case is given in [16, 36]. For demonstration purposes, we have partially modified the original requirements.

1.2.2 Dependency- and Part-of-Speech-Based Approach

One possible approach to arrive at formal requirements models is to investigate the grammar of the natural language requirements. For instance, one can make use of dependency tags, part-of-speech tags or combine both of these syntactical information to arrive at the entities for the desired requirements models.

For example, consider the requirement phrase *"if the temperature is larger than t_batt_max"*. When we parse the dependency and POS tags of this phrase with the help of toolkits like spaCy,[2] we arrive at a directed graph as shown in Fig. 1.3. It illustrates that the *root* verb of the phrase is *"is"* (from which the graph emerges) and shows how each word is associated with this *root* verb. These associations or

---

[2] https://spacy.io/, https://explosion.ai/demos/displacy.

Table 1.1 A few commonly used dependency and POS tags

Dependency tag	Description
Nsubj	Nominal subject–does the action
Root	Root of the sentence–the action
Dobj	Direct object–on which the action is done
Mark	Marker–marks a clause as subordinate to another
Conj	Conjunct–indicates a coordinating conjunction
Part-of-speech tag	Description
NOUN	Nouns denote a person, place, thing, animal or idea
ADJ	Adjectives modify nouns
ADV	Adverbs modify verbs
DET	Determiners indicate the reference of the nouns
SCONJ	Subordinating conjunction

dependencies are represented by different tags. For example, "*temperature*" is the nominal subject (*nsubj*) of the *root*, "*t_batt_max*" is the prepositional object (*pobj*) of the *root*, etc. Similarly, we can find the POS tag of each word in the graph, e.g. "*temperature*" is a *noun*, "*larger*" is an *adjective*, etc. Some dependency and POS tags are presented in Table 1.1. We suggest the reader to gather an overview of the dependency and POS tags from the generic framework presented as universal dependencies.[3]

With this background on dependencies and POS tags, we further discuss how we tailor our pipeline with a rule base depending on the syntactic structure of the require-ment text. For better comprehensibility for readers, the requirements are presented in the order of growing complexity.

> **Req. 1** The error state is 'E_BROKEN', if the temperature of the battery is larger than t_batt_max.

As an entry point to the discussion, consider a simple requirement, Req. 1. Here, the requirement has two primitive actions–one that talks about the "*error state*" and another that talks about the condition when this "*error state*" occurs. Similar to this requirement, individual industrial requirements are often composed of multi-ple actions, making them inherently difficult to process with information extraction toolkits. Hence, as shown in Fig. 1.2, we propose a decomposition step to initially decompose a long complex requirement into clauses which individually explain prim-itive actions.

---

[3] https://universaldependencies.org/u/dep/, https://universaldependencies.org/u/pos/.

Table 1.2 Detection of comparison operators using Roget's Thesaurus

Quantity	Words	Operator
Superiority	Exceed, pass, larger, greater, over, above, ...	>
Greatness	Excessive, high, extensive, big, enlarge, ...	>
Inferiority	Smaller, less, not pass, minor, be inferior, ...	<
Smallness	Below, decrease, limited, at most, no more than, ...	<
Sameness	Equal, match, reach, come to, amount to, ...	==

Decomposition (conditions). To decompose this requirement to individual clauses, we can use the conditional keyword *"if"* as the boundary. This yields us the following clauses:

- The error state is 'E_BROKEN'
- if the temperature of the battery is larger than t_batt_max

Entity detection. For the first clause, we extract the subject of the clause as *"error state"* and the object as *"E_BROKEN"*. The root verb in the clause is *"is"* which decides how to form the relation. By mapping the subject (with an inflection of the root verb) to the *signal* and the object as the *parameter*, we end up with the relation `set_error_state(E_BROKEN)`. In the second clause, with similar rules, we extract the subject of the clause as *"temperature of the battery"* and the object as *"t_batt_max"*.

Operator detection. However, the root verb occurs in conjunction with a comparative term *"larger"*. The appropriate operator for this word can be fetched with the help of a hyperlinked thesaurus like Roget's Thesaurus.[4] This yields us the symbol ">" indicating the quality "greatness", see Table 1.2. The occurrence of a comparative term differentiates how we form the relation for this clause from the previous clause. The relation formed from this clause will be: `temperature_of_battery() > t_batt_max`.

> **Req. 2** The error state is 'E_BROKEN', if the temperature of the battery is larger than t_batt_max and <u>it</u> is smaller than t_max.

---

[4] https://sites.astro.caltech.edu/~pls/roget/.

Now consider a slightly different requirement with a different formulation in the second clause as shown in Req. 2.

Decomposition (root conjunctions). In this clause, the conjunction *"and"* occurs. Unless we decompose this conjunction, the clause by itself does not address a single primitive action. In this case, the conjunction *"and"* connects the two root verbs *"is"* (before *"larger"*) and *"is"* (before *"smaller"*). Such root conjunctions can be decomposed by considering the conjunction as the boundary. This yields us the following clauses:

- if the temperature of the battery is larger than t_batt_max
- it is smaller than t_max

Pronoun resolution. The first clause is similar to the one presented in the previous requirement. However, the second clause presents a new challenge. The pronoun *"it"* needs to be resolved before proceeding to entity detection. We propose a simple pronoun resolution step to find the third person pronouns (singular: *it*, plural: *they*) by replacing each occurrence of a pronoun with the farthest subject. In this clause, we replace *"it"* with *"the temperature of the battery"*. This yields the clause *"if the temperature of the battery is smaller than t_max"*. This is again similar to the discussed clauses and is handled with the same rules. A further check on the grammatical number of the pronoun and its antecedent is advised, if the requirement quality is in doubt. The pronouns without an antecedent (called pleonastic pronouns) should not be resolved. We also assume first person or second person pronouns hardly occur in industrial requirements. For the requirements considered here, there is no particular need to use more sophisticated algorithms. If necessary, one could use, for example, the coreference pronoun resolution algorithms provided by Stanford CoreNLP[5] or AllenNLP.[6]

> **Req. 3** The error state is 'E_BROKEN', if the temperature of the battery is larger than t_batt_max <u>and</u> t_max.

Now consider a requirement which has the following clause as given in Req. 3. Unlike the clause in the previous requirement, this clause has a conjunction *"and"* between two noun phrases *"t_batt_max"* and *"t_max"*.

Decomposition (noun phrases). This demands a slightly more complex noun phrase decomposition. Unlike the other decomposition steps described above, this step further requires to extract the subject phrase of the first noun phrase and prefix it to the latter. This yields us the following clauses:

---

[5] https://stanfordnlp.github.io/CoreNLP/coref.html.

[6] https://demo.allennlp.org/coreference-resolution.

- if the temperature of the battery is larger than t_batt_max
- if the temperature of the battery is larger than t_max

These clauses are identical to the discussed clauses and hence, the entities are extracted with the same rules.

> **Req. 4** The error state is 'E_BROKEN', if the temperature of the battery is between t_batt_max and t_max.

Another clause is shown in Req. 4, with a connector "*between*" in the text.

Decomposition (connectors like "between"). To decompose such clauses, we can replace *between A and B* with the construct *greater than A and less than B*. This however, assumes the real values of *A* to be less than *B*. Although the assumption is true in most cases, we advise to further validate the decomposition, e.g. by integrating a user feedback loop. With the above assumption, the decomposition of this clause results in the following clauses:

- if the temperature of the battery is greater than t_batt_max
- if the temperature of the battery is less than t_max

> **Req. 5** The charging approval shall be given if the connection with the charging station is active.

> **Req. 6** The charging approval shall not be given if the connection with the charging station is inactive.

Requirements can also contain negations. Consider the pair of requirements Reqs. 5 and 6, which when decomposed yield the following clauses:

- *R5C1:* The charging approval shall be given
- *R5C2:* if the connection with the charging station is active

- *R6C1:* The charging approval shall not be given
- *R6C2:* if the connection with the charging station is inactive

Entity detection (negation handling). The corresponding clauses of these requirements contradict each other, i.e. *R5C1* and *R6C1* contradict each other as well as *R5C2* and *R6C2*. In *R5C1*, there is no explicit negation. Hence, we can

Table 1.3 Mapping of DEP/POS tags to syntactic entities

DEP/POS tags (constituents)	Syntactic entities
Nsubj, nsubjpass	Subject
Pobj, dobj	Object
Root, root + NOUN, auxpass/root + ADP/ADV/ADJ	Predicate

Table 1.4 Mapping of syntactic entities to model entities

Syntactic entities	Signal	Operator	Parameter
Subject, predicate	Predicate_subject	–	–
Subject, predicate, object	Predicate_subject	–	Object
ADJ, subject, predicate, object	Predicate_ADJ	–	Object
Subject, ADJ, object	Subject	Op(ADJ)	Object

handle this clause just as any other clause explained before. However, the root verb in *R6C1* *"be"* occurs in conjunction with a *"not"*. This can be addressed by introducing a negation operator (a logical not) in the relation. For example, *R5C1* yields give_charging_approval() and *R6C1* yields not give_charging_approval().

Handling the second clauses of the requirements poses a different challenge. In *R5C2*, the word *active* occurs, while in *R6C2*, the antonym of this word *inactive* occurs. We propose to assume the first occurrence of a word qualifying such boolean expressions as *boolean true* and if its antonym is cited in a different requirement, it can be assigned as *boolean false*. The antonyms are identified using the hierarchical thesaurus WordNet.[7] We assume this way of approaching negations is more intuitive than looking at a single word and inferring its sentiment. The sentiment of a word does not necessarily help us define the boolean value of the word.

In the light of the above discussion, we present an overview of a few rules we use to extract the entities in the following tables. The dependency and POS tags are first mapped to the syntactic entities *subject*, *object* and *predicate*, see Table 1.3. Then these syntactic entities are mapped to the model entities *signal* and *parameter* according to specific rules, see Table 1.4. The *operator* is identified from the ADJ denoted by op(ADJ) in the table. The rules can be extended e.g. by considering the type of action (nominal, boolean, simple, etc.) and verb types (transitive, dative, prepositional, etc.).

Discussion. Although a custom rule base exploiting dependency and POS tags is well suited for a particular use case, it demands a tremendous effort to generalize the rules for requirements across domains varying in nature and style. With each

---

[7] http://wordnetweb.princeton.edu/perl/webwn.

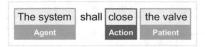

Fig. 1.4 Semantic roles for an exemplary requirement phrase

Table 1.5 List of basic semantic roles

Role	Description
Agent	Who/what performed the action
Patient	Who/what was the action performed on
Location	Where the action was performed
Time	When the action was performed
Instrument	What was used to perform the action
Beneficiary	Who/what is benefited by the action

new style of requirement, the rule base may need to be updated, leading to further effort and deliberation. We argue this by considering the number of dependency and POS tags, which are 37 dependency tags and 17 POS tags according to the revised universal dependencies [25].

1.2.3 Semantic-Role-Based Approach

We can also formalize requirements by extracting the semantic roles in the requirement text. The semantic roles represent the relation between the words or phrase in regards to the main verb present in the sentence. It describes the conceptual meaning behind the word.

For example, consider a simple requirement phrase, *"The system shall close the valve"*. As illustrated in Fig. 1.4, *"the system"* is semantically the *agent* of the action *"close"*. Further, this action is performed on *"the valve"* which is semantically its *patient*. In general, the *agent* is the one who initiates the action and the *patient* is the one on whom the action is performed. These roles are semantic properties in contrast to *subject* and *object*, we used in Sect. 1.2.2, which are syntactic properties. Some semantic roles are presented in Table 1.5.

Semantic Role Labeling (SRL) is the task of assigning these roles to the phrases or words that describe their semantic meaning. SRL is one of the leading tasks in natural language processing [26]. It can be used for applications such as question answering, data mining etc. We use SRL for requirements formalization by identifying the roles to form the desired requirements models. We use a state-of-the-art pre-trained deep learning BERT encoder with BiLSTM [35] by AllenNLP[8] which was trained on CoNLL 2005 and 2012 datasets. This model generates frames, each containing

---

[8] https://demo.allennlp.org/semantic-role-labeling.

Table 1.6 Basic arguments based on English Propbank

	Argument	Role
Numbered	ARG0	Agent
	ARG1	Patient
	ARG2	Instrument/beneficiary
	ARG3	Start point/beneficiary
	ARG4	End point
Functional	ARGM-LOC	Location
	ARGM-TMP	Time
	ARGM-NEG	Negation
	ARGM-PRD	State

a set of arguments related to a verb in the sentence. The arguments are defined based on English Propbank defined by Bonial et al. [6]. Each argument describes the semantic relation of a word to the verb with which it is associated. The arguments can be either numbered or functional. Table 1.6 shows some of the arguments and the corresponding roles.

We will now demonstrate how we handle the decomposition and entity detection stages with the SRL roles. This can help in considerably reducing the number of rules compared to the first approach while handling even more complex requirements. It is worth to note that there are only 24 SRL arguments (5 numbered and 19 functional) and we argue that even an exhaustive rule base would only demand a combination of some of these different arguments.

> **Req. 7** The maximum power shall be limited to [G_Max] and the event "High device temperature" shall be indicated when the device temperature exceeds [T_Hi] °C.

Consider the requirement Req. 7. This requirement has three desired clauses with two statements based on one condition.

Preprocessing. In the requirement, we have the name of an event *"High device temperature"* within quotes which describes what happens when this event occurs in the system. In certain other cases, the use case has longer event names which by itself had semantic roles within. To help the SRL model to distinguish the event names from the rest of the requirement text, we preprocess the requirement by replacing the occurrences of event names in quotes with abbreviated notations like *"E1"*. We also eliminate square brackets and units around variables, for example *"G_Max"* and *"T_Hi"* in Req. 7.

The pre-trained SRL model also retrieves frames for modal verbs like *"be"* or *"shall"* which may not have any related arguments. To devise a rule set that works with almost all frames, we discard frames with just one argument.

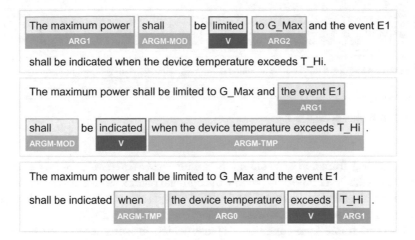

Fig. 1.5 SRL frames for Req. 7

Decomposition. Unlike the decomposition described in Sect. 1.2.2 which utilized dependency and POS tags, here we discuss the decomposition of complex requirements based on the detected SRL frames. As shown in Fig. 1.5, we obtain a total of three frames from the pre-trained SRL model applied on Req. 7. The span (obtained by adding up the requirement phrases belonging to each role in the frame) of these detected frames yield us the following decomposed clauses:

- the maximum power shall limited to G_Max
- the event E1 shall indicated
- when the device temperature exceeds T_Hi

In the second clause, we have the role *ARGM-TMP* which describes the condition of the action. Here, we have *ARGM-TMP* with the desired conditional clause as its span which tells us that this condition is related to this particular action. But when we look at the requirement we know that the condition clause applies to both the other clauses. So as to avoid wrong formation of the output model, we will avoid those *ARGM-TMP* role with the full clause as span and only consider those *ARGM-TMP* roles with a single word like *"when"* or with the unit of time.

Entity detection. The first frame is associated with the verb *"limited"*. We have *"The maximum power"* as *ARG1* (describing on whom the action is performed on), *"to G_Max"* as *ARG2* (an instrument). We can map the verb together with *ARG1* as the signal and *ARG2* as the parameter. Note that we use the lemma (basic form) of the verb to form the signal. Also, the stop words in the arguments are eliminated. This yields us the relation for this clause as: limit_maximum_power(G_Max). From this relation, we form the construct V_ARG1(ARG2) which can be further used if we get similar frames.

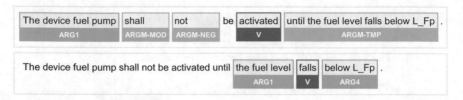

Fig. 1.6 SRL frames for Req. 8

The second frame, after avoiding the argument *ARGM-TMP*, follows a similar construct as the first frame only without the optional parameter. We also perform back mapping of the abbreviated event name *"E1"* with its actual event name yielding the relation: `indicate_event_high_device_temperature()`.

The third frame is a conditional clause which is identified by the argument *ARGM-TMP* with a span of *"when"*. Here, we have *"the device temperature"* as *ARG0* (the agent) and *"T_Hi"* as *ARG1* (the patient). We map *ARG0* as the signal and *ARG1* as the parameter.

Operator detection. Additionally, a comparison word *"exceeds"* occurs in the role *V* of this clause. We detect the corresponding operator by looking it up in Roget's Thesaurus, similar to the first approach. In this case, the word *"exceeds"* gives the symbol *">"*. Thus, this clause translates to the relation `device_temperature() > T_Hi`. We build a general construct for similar frames with operator as `ARG0 V ARG1`.

Req. 8 The device fuel pump shall <u>not</u> be activated until the fuel level falls below [L_Fp].

Consider another requirement, Req. 8, with negation and the conditional keyword *"until"*. SRL frames obtained from the pre-trained model are shown in Fig. 1.6.

Entity detection (negation handling). The first frame follows the same construct as before leaving the argument *ARG2* optional. However, we also have the argument *ARGM-NEG* indicating a negation in the clause which should be considered while forming the relation. In this case, we can modify the previous construct `V_ARG1()` to form a new construct `not V_ARG1()`. This finally yields the relation `not activate_device_fuel_pump()`.

Though the second frame indicates a conditional clause, it is difficult to identify since no arguments have been assigned to the word *"until"*. So as to recognize it as a condition, we apply rules to identify it as condition like keyword identification. Considering the arguments, we have the role *ARG4*, which indicates an ending point that can be mapped to a parameter and *ARG1* to a signal.

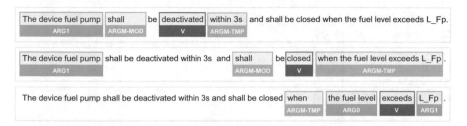

Fig. 1.7 SRL frames for Req. 9

Operator detection. To find the operator symbol, the verb text will not be enough in this case as *"falls"* alone cannot help in getting the operator symbol. So, as to get the symbol, we apply one extra rule, i.e., to consider the next word in the requirement in addition to the verb text to get the symbol. So the span text would be *"falls below"* which gives the symbol "<".

So this particular frame would lead to `fuel_level() < L_Fp` leading to the formation of the construct `ARG1 V ARG4`.

> **Req. 9** The device fuel pump shall be deactivated within 3s and shall be closed when the fuel level exceeds [L_Fp].

SRL can handle the time constraint and some complex decomposition as well. To demonstrate this, we will modify the previous requirement as shown in Req. 9. Figure 1.7 shows the identified SRL frames.

Decomposition. When we decompose this requirement with keywords like *"and"* and *"when"*, it would lead to wrong formation of clause as the second clause would not show what shall be closed. Looking at the SRL frames, it correctly identifies this and forms the correct following clause by considering span text for each frame.

- The device fuel pump shall deactivated within 3s
- The device fuel pump shall closed
- when the fuel level exceeds L_Fp

The first and the second frame follows the same construct as before, i.e., `V_ARG1()`, and the third frame follows `ARG0 V ARG4`. In the first frame, we have identified *ARGM-TMP* with a time constraint. This temporal behavior can be detected by some rule, e.g. whether the span text contains a number and a unit of time, or from the keyword *"within"*. However, we leave the discussion of modeling non-functional properties such as temporal behavior to future work.

In Table 1.7 we have summarized some rules for mapping the arguments to model entities. The first three lines in the table show some cases for assignments, whereas the two last lines show cases for conditions. Conditions are identified using

Table 1.7 Mapping of SRL arguments to model entities

SRL arguments	Signal	Operator	Parameter
ARG1 (patient), V	V_ARG1	–	–
ARG1 (patient), V, ARG2 (instrument)	V_ARG1	–	ARG2
ARG1 (patient), V, ARGM-PRD (state)	V_ARG1	–	ARGM-PRD
ARG0 (agent), V, ARG1 (patient)	ARG0	op(V)	ARG1
ARG1 (agent), V, ARG4 (end points)	ARG1	op(V)	ARG4

the argument *ARGM-TMP*. The *operator* is identified from the *V* role denoted by op(V) in the table.

Discussion. Using SRL information, we need much easier and less rules as compared to the first approach. This makes it easy to adapt this approach according to a specific use case and needs. However, in a few cases we found that SRL was not working properly. For further improvements, one could of course combine both approaches using the POS and dependency tags as well as SRL labels.

Similar as in the first approach, the underlying deep learning models are evolving quickly. They are trained with new and better annotated datasets, resulting in better generation of models with state-of-the-art performance and higher accuracy. The above mentioned constructs and rules are based on the specific model (current state of the art [35]) we used to extract SRL arguments. This might not work completely with the future better models as they would extract more information and these constructs/rules could be outdated. So it might be necessary to change the rules when other models are used.

1.2.4 Model Formation

Once the necessary entities are extracted from the requirement clauses, we head to the last stage in the pipeline, model formation. As mentioned above, we form a relation for each clause. In particular, assignments are of the form `signal (parameter*)` with optional parameter, and conditional statements are of the form `signal()` `operator parameter`. To generate requirement models from these relations, we use a simple domain-specific language (DSL) with abstract logical blocks. We extend the modalities of our previous work [16] following the Temporal Action

Language (TeAL) of Li et al. [21]. It maps the relations according to some identified keywords to blocks. Conditional statements starting with *if/when/while* are mapped to an `if`-block, those starting with *until* to an `until`-block. Assignments without introductory keyword are mapped to a `then`-block and those starting with *else/otherwise* to an `else`-block. When there is no conditional statement given in the requirement, we map all assignments to a statement-block without using any keywords. Conjunctions (*and/or*) identified between relations are also accommodated in these DSL blocks. Though this mapping appears rather trivial, our aim is to make this translation simple and flexible, so that ways are open for integration with other sophisticated languages. Obviously, our DSL is very similar to the FRETISH language [14]. The resulting models can also be further transformed into Matlab Simulink models or UML sequence diagrams, depending on what the end user desires.

For example, the model for Req. 1 is of the form:

```
if( temperature_of_battery() > t_batt_max )
    then( set_error_state(E_BROKEN) )
```

Similarly, the more complex Req. 7 will yield the following model:

```
if( device_temperature() > T_Hi )
    then( limit_maximum_power(G_Max)
        and indicate_event_high_device_temperature())
```

In the same way `else`-blocks and `until`-blocks are represented.

Discussion. There exist many different, more or less common textual and graphical representations of behavioral models, for example temporal logic, cause-effect graphs, UML diagrams (sequence diagrams, class diagrams, activity diagrams, etc.), OCL constraints, Petri nets, or Matlab Simulink models. Initially, we planned to use sequence diagrams as output format since it is very common and can be represented textually and graphically.[9] However, we realized that some parts of the requirements, such as *until*-conditions or temporal behavior, cannot be handled easily. For more complex requirements, the overview gets lost due to many nested fragments. Pseudocode has the advantage of being human- and machine-readable. It is a kind of more abstract, intermediate format, one has to define a mapping for the keywords for further processing. We see this as an advantage, since this mapping can be adapted to different use cases and needs. It includes all information to be mapped to other model formats.

---

[9] https://sequencediagram.org/.

1.3 Existing Methodologies

In this section, we want to give a brief overview of existing methodologies for requirements formalization.

1.3.1 Controlled Natural Language (CNL)

The use of a Controlled Natural Language (CNL) is an old and common approach to enable automatic analysis. The requirements need to be written in restricted vocabulary and following some specific templates (also called patterns or boilerplates). Commonly used templates include the SOPHIST template [33], the EARS template [27] and the FRETISH template [14].

Early NLP methods used dictionaries and a large set of rules to extract syntactic and semantic information such as syntax trees and semantic roles to form models [8, 32]. More recently, Allala et al. [2] exploit templates to represent requirements and utilize Stanford CoreNLP to create test cases and meta-modeling for validation. The FRET tool can be used to map the structured requirement with the help of templates and convert them into formal models [11]. Giannakopoulou et al. [14] use FRETISH requirements and derive temporal logic formulas from them. However, these approaches use rule sets rather than any kind of NLP method.

1.3.2 Part-of-Speech Tagging and Dependency Parsing

As explained in Sect. 1.2.2, part-of-speech (POS) tags categorize words according to their grammatical properties, and dependency tags identify grammatical structure of the sentence. Different libraries like NLTK, spaCy and Stanford CoreNLP can be used for POS tagging and dependency parsing. Dependency annotation can also be done using Universal Dependencies Guidelines on software requirements [18].

Pre-defined patterns are considered on dependency trees to identify subtrees which would help in generation of cause-effect-graphs [12]. Unstructured Information Management Architecture (UIMA) framework has been used to extract POS tags and determine phrases to help identify actions and conditions to form models on use case descriptions [37]. Fischbach et al. [12] also construct few patterns with combination of POS and dependency tags. Koscinski et al. [20] use Stanford CoreNLP to assign dependency tags and use pre-defined patterns to identify syntactic categories from which requirements models can be generated with a small amount of rules. Fritz et al. [13] uses the combination of both POS and dependency tags to identify semantic frames to map requirements into an existing structure. The latter two papers are very similar to our first approach, but use different categories and rules.

1.3.3 Named Entity Recognition (NER)

Phrases from sentences can be classified into some predefined categories known as named entities. NER is the task of assigning these entities. Commonly used categories are person, organization, location etc. But these categories are not helpful for requirements analysis. To perform NER for requirement texts, one needs to define new categories and manually annotate a dataset to train ML-based approaches. Annotating a dataset is a laborious and a time consuming task. Also, defining categories can vary depending on the requirement type. Malik et al. [24] and Herwanto et al. [19] defined 10 and 5 entities, respectively, for software requirement specifications and privacy related tasks. In both approaches, the dataset is annotated manually. Multiple checks with multiple human annotators are carried out to get better quality data. Nayak et al. [28] have trained an NER model for creating expressions with 9 entity categories, which is then used to formalize requirements.

1.3.4 Semantic Role Labeling (SRL)

As described in Sect. 1.2.3, in Semantic Role Labeling (SRL), words or phrases in a sentence are assigned labels that indicate their semantic role in the sentence, such as the agent, the patient, etc. Semantic frames have been widely used for requirements analysis in recent years. From simple dependency parsing rules [34] to machine learning models [4, 35], much has been used to extract semantic roles and frames. Sengupta et al. [34] use the Stanford Dependency manual to create rules that help extract basic arguments for requirement analysis, whereas Fritz et al. [13] use spaCy to extract POS and dependency tags from which they create rules to extract semantic roles. Rules can also be created to recognize semantic relations based on the lexeme (word form) and its variants [5].

FrameNet was utilized to determine semantic relationships between requirements with inheritance knowledge [1]. VerbNet and other open source resources were used to perform semantic analysis for specification representation [23].

Recently, machine learning techniques have also been considered for semantic role extraction. Wang et al. [39] use the SRL model of the CogComp NLP pipeline to generate domain models with a rule set. Mate Tools [4] has been a commonly used ML model for SRL which uses linear logistic regression. Diamantopoulos et al. [10] extended Mate tools' semantic role labeler by training it with additional lexical attributes as features, and used it for mapping of requirements to formal representations. The latter work is similar to our second approach but we utilize a much more recent SRL model with very flexible rules.

1.3.5 Translation and Back-Translation

Translation of requirement texts to some kind of model directly in one step is not an easy task. As seen from all the above mentioned methods, it requires multiple steps and multiple methods to form the output model. We can use the concept of Neural Machine Translation (NMT)/Code Generation to translate requirements to desired pseudocode. Code generation has been used to convert complex text into programming language code using machine learning models [22]. It can also be used to convert text into an Abstract Syntax Tree (AST) which represents syntactic structure of the desired code [30]. This would require manual creation of a large dataset with requirement texts and its corresponding pseudocode or AST which is again a laborious and a time consuming task.

Neural Machine Translation can also be used to back translate output models to requirement text. Tools like OpenNMT has been used to convert temporal logic into natural language requirements [9]. A set of rules can also be applied to create functional natural language requirements from data models [7]. A systematic survey on the generation of textual requirements specifications from models can be found in [29].

1.4 Discussion

In view of the many recent developments in this area of research, the journey towards the development of mature requirements formalization techniques has only just begun. Since the aim of this work is to present principal ideas on this research topic, we also want to discuss some findings and shortcomings that we identified during our investigations. The tremendous survey of Zhao et al. [40] has already identified some key findings for the NLP-supported requirements engineering process, namely (i) a huge gap between the state of the art and the state of the practice, i.e., an insufficient industrial evaluation, (ii) little evidence of industrial adoption of the proposed tools, (iii) the lack of shared RE-specific language resources, and (iv) the lack of NLP expertise in RE research to advise on the choice of NLP technologies. We agree that these findings also apply to the requirements formalization process.

1.4.1 Lessons Learned

The international XIVT project[10] addressed the challenge of automatically analyzing and extracting requirements in order to minimize the time for testing highly configurable, variant-rich systems. During the three-year project period, we had many interesting talks and discussions with research and industry partners on this

---

[10] https://www.xivt.org/.

research area. The approaches presented in this work are the result of countless internal and external discussions, which shows that it is not an easy task and very manifold, but also a very interesting topic.

Generality. It is very tempting to try to develop a single algorithm that is general enough to handle a wide range of natural language requirements. But, just as natural language is diverse, so would the algorithm to be developed have to be. Requirements are formulated by such a wide variety of people with different motivations, experience and expert knowledge, from different branches and with different social backgrounds, that it is currently impossible to create one algorithm for the huge variety of styles of unstructured natural language. The many existing approaches and our own developments show that it is reasonable to automate this process to a high degree for a specific use case, domain and writing style. In short: *Don't worry about being specific!*

Requirement quality. The human language is full of misunderstandings, ambiguities, inconsistent and incomplete formulations. When developing a rule set we were strongly biased by the given requirements. We kept trying to improve the rules to make them more and more accurate. In discussion with our industry partners, we realized that some parts of the requirements seemed to be somewhat poorly written. By a short reformulation, our algorithm could handle them much easier. This shows that the algorithmic treatment of the formalization process reveals many minor issues that need to be fixed by the requirement engineers, and thus also contributes to a higher quality of the requirements. In short: *Don't always blame the model!*

Degree of automation. For a long time of our development, we planned to build a standard input/output tool that would generate the requirements models fully automatically without asking the user for any input or confirmation. However, we have found that we are unable to properly process the data if some information within the pipeline is missing. Additionally, our industry partners have also requested to always have a quick verification option by prompting missing or uncertain information to the user. So if even humans need to review many details to formalize a set of requirements, how should an algorithm be able to generate correct and complete models fully automatically? The fact that requirements analysis is a highly interactive process should also be taken into account when implementing the tool. It is much more desirable from the industry to have assistance in creating correct requirements models than to have a fully automated algorithm that produces some erroneous models. In the end, the required working time of an engineer decides whether the review of automatically generated models is preferable to the manual creation process. That is, the algorithms need to be highly efficient to have the chance of being used by the industry, therefore we believe that an interactive, performant user interface is essential. In addition, the information obtained from user input can also be used to improve the underlying ML algorithms. In short: *Go for user interaction!*

Availability of data. A crucial obstacle in this research area is that the data situation for freely available requirements with corresponding models is very poor. One would need a large corpus of these data from different use cases to compare different algorithms. Unless a reliable and sufficiently large amount of ground truth exists, a benchmark of different approaches and tools is not possible [38]. Even if one takes the time in a group of scientists to analyze, label, and model requirements, detailed discussions with industry stakeholders are essential and time-consuming. Consider the quote of Steve Jobs: *"A lot of times, people don't know what they want until you show it to them"* [31]. Thus, the success of ML-trained algorithms relies on having a large amount of labeled and unlabeled data sets available. In short: *Share your data!*

Availability of tools. Similarly, the availability of ready-to-use algorithms and tools is very limited. Only a few tools are publicly available [40], and most of them are in a prototypical stage. This makes it very difficult for industry to identify useful tools for the application to their use cases. In short: *Go for GitHub*[11]*!*

Evolution. Most approaches in the literature deal with a fixed set of requirements. Likewise, most freely available requirements specifications are provided in a single fixed document (with a few exceptions[12]). However, this is completely contrary to the highly iterative and evolving process of requirements elicitation. Just as dynamic as software development is the corresponding requirements engineering. On the other hand, we observe that NLP models have evolved very rapidly in recent years. Since the publication of the BERT models, there has been an exploding amount of research about the application and further training of these models. Researchers are hunting for better and better models with higher accuracy, algorithms from one or two years ago are already outdated and need to be revised when used for further development. In short: *Everything evolves!*

1.4.2   Possible Future Developments

From our perspective, there are a number of emerging, sophisticated NLP methods and models that can be applied to the requirements formalization process. Although it is very difficult to look into the distant future of NLP research and predict developments, we do see some direct follow up to current approaches and can envision some future developments.

Named entity recognition (NER). The advantage of NER is the direct identification of model entities when they are used as labels. However, the labeled dataset must be very large to achieve reasonable accuracy. To reduce the intensive labelling work, one could use pre-trained models (e.g. provided by AllenNLP) and fine-tune them with newly annotated datasets with new categories.

---

[11] https://github.com/, or similar platforms.

[12] AUTOSAR, https://www.autosar.org/nc/document-search/.

Semantic role labeling (SRL). The current SRL models already have an acceptable accuracy. However, we also found that the predictions can change suddenly if just one word or comma is changed. The model may not work as well for domain-specific wording or the writing style of requirements. Therefore, the improvement in accuracy could be studied if the SRL models are pre-trained with some labeled requirements data (a starting point could be [10].

Question-answer driven semantic role labeling (QA-SRL). As mentioned above, requirements analysis is highly interactive and needs several input from the user. It would be a new level of automation, if an algorithm itself is able to formulate questions and process the given answers to clarify missing or uncertain information for generating the model.

Transfer learning (TL). It is very successful in other fields of NLP research to train a model on some related tasks with a larger amount of data and then to fine-tune it for the actual task with much less labeled data. We could imagine using models that have already been trained for requirements classification or ambiguity detection.

Online learning. Current approaches use NLP models that are trained once and will not change according to user feedback. It would be much more convenient for the user if the algorithm learns from the input and suggests better results for subsequent requirements. This could be done by some sort of post-training of the models or more sophisticated continuous learning methods. This could also be useful for a completely new use case and domain, where the algorithm learns the rules from the beginning and one could avoid implementing a rule set at all.

Translation/back-translation. Directly translating requirements into models with a large ML model might be far in the future or somehow not reasonable. But there are interesting developments in the research area of translating text to code that could be helpful in formalizing requirements as well. The other direction of generating boilerplates sounds much simpler, but also needs to be studied in detail. We can imagine that in the future a coexistence of requirements and models will be possible, where hand-written requirements are transformed into models and transformed back into structured text that can be used for further processing–a kind of "speaking models".

Semantic similarity. As shown in our previous work [16], semantic similarity between requirements and product design descriptions (if available) is helpful for generating more concrete requirements models, i.e., identifying signal and parameter names from the design specification instead of generating abstract entity names. However, identifying the correct Boolean value (antonym detection), for example, is still a difficult task in NLP research. Moreover, requirements formalization and quality analysis are strongly interlinked. Semantic similarity can be useful for consistency checking, e.g., to identify identical signals and parameters used with slightly different terms within the set of requirements, see [39] for a first approach.

Auxiliary documents. Requirements do not fall from the sky. Usually, an extensive elicitation phase takes place in advance. One could make use of preparatory documents such as interviews, meeting notes, etc., or implementation artifacts such as issue descriptions (bug reports, feature requests, etc.). They can provide context, different formulations, and missing information. It would also be interesting to explore how to support the entire requirements development process. One would need to study the process in great detail, track and analyze the documents, and identify some successful workflows that can be assisted by NLP approaches (a starting point could be the FRET tool [14]).

1.5 Conclusion

In this work, we have presented and discussed principal ideas and state-of-the-art methodologies from the field of NLP to automatically generate requirements models from natural language requirements. We have presented an NLP pipeline and demonstrated the iterative development of rules based on NLP information in order to guide readers on how to create their own set of rules and methods according to their specific use cases and needs. We have studied two different approaches in detail. The first approach, using dependency and POS tags, shows good results but is somewhat inflexible in adopting the underlying rule set for different use cases and domains. The second approach, using semantic roles, shows similar good results, but requires less effort to create a set of rules and can be easily adapted to specific use cases and domains. The use of a human- and machine-readable format for the requirements models appears suitable for easy reading comprehension and at the same time for automatic further processing. The results show that the current pre-trained NLP models are suitable to automate the requirements formalization process to a high degree.

Furthermore, we provided an overview of the literature and recent developments in this area of research and discussed some findings and shortcomings (lessons learned) that we identified throughout the development of our approaches. Finally, we suggested some possible future developments in NLP research for an improved requirements formalization process.

Acknowledgements This research was funded by the German Federal Ministry of Education and Research (BMBF) within the ITEA projects XIVT (grant no. 01IS18059E) and SmartDelta (grant no. 01IS21083E). We thank AKKA Germany GmbH and Alstom for providing industrial use cases for the demonstration of the presented methods. We are grateful to Prof. Holger Schlingloff from Fraunhofer FOKUS/HU Berlin for his valuable recommendations.

References

1. Alhoshan, W., Zhao, L., Batista-Navarro, R.: Using semantic frames to identify related textual requirements: An initial validation. In: ESEM '18, pp. 1–2. ACM (2018). https://doi.org/10.1145/3239235.3267441
2. Allala, S.C., Sotomayor, J.P., Santiago, D., King, T.M., Clarke, P.J.: Towards transforming user requirements to test cases using MDE and NLP. In: COMPSAC 2019, pp. 350–355. IEEE (2019). https://doi.org/10.1109/COMPSAC.2019.10231
3. Ammann, P., Offutt, J.: Introduction to Software Testing, 2nd edn. Cambridge University Press (2017). https://doi.org/10.1017/9781316771273
4. Björkelund, A., Hafdell, L., Nugues, P.: Multilingual semantic role labeling. In: CoNLL 2009, pp. 43–48. ACL (2009). https://aclanthology.org/W09-1206
5. Bokaei Hosseini, M., Breaux, T.D., Niu, J.: Inferring ontology fragments from semantic role typing of lexical variants. In: REFSQ 2018, LNCS, vol. 10753, pp. 39–56. Springer (2018). https://doi.org/10.1007/978-3-319-77243-1_3
6. Bonial, C., Bonn, J., Conger, K., Hwang, J.D., Palmer, M.: PropBank: Semantics of new predicate types. In: LREC 2014, pp. 3013–3019 (2014). https://aclanthology.org/L14-1011/
7. de Brock, B., Suurmond, C.: NLG4RE: How NL generation can support validation in RE. In: NLP4RE'22, CEUR-WS.org (2022). http://ceur-ws.org/Vol-3122/NLP4RE-paper-2.pdf
8. Carvalho, G., Barros, F., Carvalho, A., Cavalcanti, A., Mota, A., Sampaio, A.: NAT2TEST tool: From natural language requirements to test cases based on CSP. In: SEFM 2015, LNPSE, vol. 9276, pp. 283–290. Springer (2015). https://doi.org/10.1007/978-3-319-22969-0_20
9. Cherukuri, H., Ferrari, A., Spoletini, P.: Towards explainable formal methods: From LTL to natural language with neural machine translation. In: REFSQ 2022, LNCS, vol. 13216, pp. 79–86. Springer (2022). https://doi.org/10.1007/978-3-030-98464-9_7
10. Diamantopoulos, T., Roth, M., Symeonidis, A., Klein, E.: Software requirements as an application domain for natural language processing. Lang. Resour. Eval. **51**(2), 495–524 (2017). https://doi.org/10.1007/s10579-017-9381-z
11. Farrell, M., Luckcuck, M., Sheridan, O., Monahan, R.: FRETting about requirements: Formalised requirements for an aircraft engine controller. In: REFSQ 2022, LNCS, vol. 13216, pp. 96–111. Springer (2022). https://doi.org/10.1007/978-3-030-98464-9_9
12. Fischbach, J., Vogelsang, A., Spies, D., Wehrle, A., Junker, M., Freudenstein, D.: SPECMATE: Automated creation of test cases from acceptance criteria. In: ICST 2020, pp. 321–331. IEEE (2020). https://doi.org/10.1109/ICST46399.2020.00040
13. Fritz, S., Srikanthan, V., Arbai, R., Sun, C., Ovtcharova, J., Wicaksono, H.: Automatic information extraction from text-based requirements. Int. J. Knowl. Eng. **7**(1), 8–13 (2021). https://doi.org/10.18178/ijke.2021.7.1.134
14. Giannakopoulou, D., Pressburger, T., Mavridou, A., Schumann, J.: Generation of formal requirements from structured natural language. In: REFSQ 2020, LNCS, vol. 12045, pp. 19–35. Springer (2020). https://doi.org/10.1007/978-3-030-44429-7_2
15. Gröpler, R., Sudhi, V., Calleja García, E.J., Bergmann, A.: NLP-based requirements formalization for automatic test case generation. In: CS&P'21, CEUR-WS.org, pp. 18–30 (2021). http://ceur-ws.org/Vol-2951/paper15.pdf
16. Gröpler, R., Sudhi, V., Kutty, L., Smalley, D.: Automated requirement formalization using product design specifications. In: NLP4RE'22, CEUR-WS.org (2022). http://ceur-ws.org/Vol-3122/NLP4RE-paper-1.pdf
17. Grujic, D., Henning, T., García, E.J.C., Bergmann, A.: Testing a battery management system via criticality-based rare event simulation. In: MSCPES'21, pp. 1–7. ACM (2021). https://doi.org/10.1145/3470481.3472701
18. Hassert, N., Ménard, P.A., Galy, E.: UD on software requirements: Application and challenges. In: UDW 2021, pp. 62–74. ACL (2021). https://aclanthology.org/2021.udw-1.5/
19. Herwanto, G.B., Quirchmayr, G., Tjoa, A.M.: A named entity recognition based approach for privacy requirements engineering. In: REW 2021, pp. 406–411. IEEE (2021). https://doi.org/10.1109/REW53955.2021.00072

20. Koscinski, V., Gambardella, C., Gerstner, E., Zappavigna, M., Cassetti, J., Mirakhorli, M.: A natural language processing technique for formalization of systems requirement specifications. In: REW 2021, pp. 350–356. IEEE (2021). https://doi.org/10.1109/REW53955.2021.00062
21. Li, W., Hayes, J.H., Truszczyński, M.: Towards more efficient requirements formalization: A study. In: REFSQ 2015, LNCS, vol. 9013, pp. 181–197. Springer (2015). https://doi.org/10.1007/978-3-319-16101-3_12
22. Ling, W., Blunsom, P., Grefenstette, E., Hermann, K.M., Kočiský, T., Wang, F., Senior, A.: Latent predictor networks for code generation. In: ACL 2016, pp. 599–609 (2016). https://doi.org/10.18653/v1/P16-1057
23. Mahmud, N., Seceleanu, C., Ljungkrantz, O.: Specification and semantic analysis of embedded systems requirements: From description logic to temporal logic. In: SEFM 2017, LNCS, vol. 10469, pp. 332–348. Springer (2017). https://doi.org/10.1007/978-3-319-66197-1_21
24. Malik, G., Cevik, M., Khedr, Y., Parikh, D., Başar, A.: Named entity recognition on software requirements specification documents. In: Canadian AI 2021. CAIAC (2021). https://doi.org/10.21428/594757db.507c7951
25. de Marneffe, M.C., Manning, C.D., Nivre, J., Zeman, D.: Universal dependencies. Comput. Linguist. 1–54 (2021). https://doi.org/10.1162/coli_a_00402
26. Màrquez, L., Carreras, X., Litkowski, K.C., Stevenson, S.: Semantic role labeling: An introduction to the special issue. Comput. Linguist. **34**(2), 145–159 (2008). https://doi.org/10.1162/coli.2008.34.2.145
27. Mavin, A., Wilkinson, P., Harwood, A., Novak, M.: Easy approach to requirements syntax (EARS). In: RE '09, pp. 317–322. IEEE (2009). https://doi.org/10.1109/RE.2009.9
28. Nayak, A., Timmapathini, H.P., Murali, V., Ponnalagu, K., Venkoparao, V.G., Post, A.: Req2Spec: Transforming software requirements into formal specifications using natural language processing. In: REFSQ 2022, LNCS, vol. 13216, pp. 87–95. Springer (2022). https://doi.org/10.1007/978-3-030-98464-9_8
29. Nicolás, J., Toval, A.: On the generation of requirements specifications from software engineering models: A systematic literature review. Inf. Softw. Technol. **51**(9), 1291–1307 (2009). https://doi.org/10.1016/j.infsof.2009.04.001
30. Rabinovich, M., Stern, M., Klein, D.: Abstract syntax networks for code generation and semantic parsing. In: ACL 2017, pp. 1139–1149 (2017). https://doi.org/10.18653/v1/P17-1105
31. Ratcliffe, S. (ed.): Oxford Essential Quotations, 6th edn. Oxford reference online. Oxford University Press (2018). https://doi.org/10.1093/acref/9780191866692.001.0001
32. Riebisch, M., Hübner, M.: Traceability-driven model refinement for test case generation. In: ECBS'05, pp. 113–120. IEEE (2005). https://doi.org/10.1109/ECBS.2005.72
33. Rupp, C., SOPHISTen: Requirements-Engineering und -Management, 7th edn. Hanser (2021). https://doi.org/10.3139/9783446464308
34. Sengupta, S., Ramnani, R.R., Das, S., Chandran, A.: Verb-based semantic modelling and analysis of textual requirements. In: ISEC '15, pp. 30–39. ACM (2015). https://doi.org/10.1145/2723742.2723745
35. Shi, P., Lin, J.: Simple BERT models for relation extraction and semantic role labeling (2019). https://arxiv.org/abs/1904.05255. Pre-print
36. Simonson, E.: An MBD adoption story from Bombardier Transportation. In: MATLAB EXPO 2018 Sweden (2018). https://www.matlabexpo.com/se/2018/proceedings/
37. Sinha, A., Paradkar, A., Kumanan, P., Boguraev, B.: A linguistic analysis engine for natural language use case description and its application to dependability analysis in industrial use cases. In: DSN 2009, pp. 327–336. IEEE (2009). https://doi.org/10.1109/DSN.2009.5270320
38. Tichy, W.F., Landhäußer, M., Körner, S.J.: nlrpBENCH: A benchmark for natural language requirements processing. In: SE 2015, pp. 159–164. GI (2015). https://dl.gi.de/handle/20.500.12116/2542

39. Wang, C., Pastore, F., Goknil, A., Briand, L.C.: Automatic generation of acceptance test cases from use case specifications: An NLP-based approach. IEEE Trans. Softw. Eng. **48**(2), 585–616 (2022). https://doi.org/10.1109/tse.2020.2998503
40. Zhao, L., Alhoshan, W., Ferrari, A., Letsholo, K.J., Ajagbe, M.A., Chioasca, E.V., Batista-Navarro, R.T.: Natural language processing for requirements engineering: A systematic mapping study. ACM Comput. Surv. **54**(3), 1–41 (2021). https://doi.org/10.1145/3444689

Chapter 2
Left Recursion by Recursive Ascent

Roman R. Redziejowski

Abstract Recursive-descent parsers can not handle left recursion, and several solutions to this problem have been suggested. This paper presents yet another solution. The idea is to modify recursive-descent parser so that it reconstructs left-recursive portions of syntax tree bottom-up, by "recursive ascent".

2.1 Introduction

Recursive-descent parser is a collection of "parsing procedures" that call each other recursively, mimicking the derivation process. This simple scheme cannot be applied if the grammar is left-recursive: a parsing procedure may indefinitely call itself, every time facing the same input. Some solutions have been suggested, and they are outlined in Sect. 2.9 at the end. We present here yet another approach. We view parsing as the process of reconstructing the syntax tree of given string. For this purpose, we equip parsing procedures with "semantic actions" that explicitly produce that tree. Recursive descent reconstructs syntax tree starting from the top. We suggest to reconstruct left-recursive portions of the tree bottom-up, by a process that we call "recursive ascent". This is also done by procedures that call each other recursively and have their own semantic actions. These procedures can be regarded as new parsing procedures. We incorporate them into the recursive-descent parser, so that recursive ascent is at the end carried out as part of recursive-descent parser for a new grammar, referred to as the "dual grammar". The dual grammar is not left-recursive, and its parser produces syntax tree for the original grammar.

In Sect. 2.2 we recall the necessary notions: BNF grammar, derivation, and syntax tree. In Sect. 2.3 we recall the idea of recursive-descent parser and introduce "semantic actions" that reconstruct the syntax tree. After introducing the necessary concepts in Sect. 2.4, we introduce in Sect. 2.5 the idea of recursive ascent together

R. R. Redziejowski (✉)
Ceremonimästarvägen 10, 181 40 Lidingö, Sweden
e-mail: roman@redz.se
URL: https://www.romanredz.se

© The Author(s), under exclusive license to Springer Nature Switzerland AG 2023   29
B. Schlingloff et al. (eds.), *Concurrency, Specification and Programming, Studies in Computational Intelligence 1091*, https://doi.org/10.1007/978-3-031-26651-5_2

with procedures to perform it. They are illustrated by two examples in Sect. 2.6. In Sect. 2.7 we point out that the procedures just defined are parsing procedures for dual grammar. Two Propositions state the essential properties of that grammar. Section 2.8 looks at implementation of choice expressions, Sect. 2.9 outlines other known solutions. Some unsolved problems are discussed in Sect. 2.10. Proofs of the Propositions are given in the Appendix.

2.2 The Grammar

We consider a BNF-like grammar $G = (\mathbb{N}, \Sigma, \mathbb{E}, \mathcal{E}, N_s)$ with set $\mathbb{N}$ of *non-terminals*, set Σ of *terminals*, set $\mathbb{E}$ of *expressions*, function $\mathcal{E}$ from non-terminals to expressions, and *start symbol* $N_s \in \mathbb{N}$.

An expression is one of these:

- $a \in \Sigma$ ("terminal"),
- $N \in \mathbb{N}$ ("non-terminal"),
- $e_1 \ldots e_n$ ("sequence"),
- $e_1 | \ldots | e_n$ ("choice"),

where each of e_i is an expression. The function $\mathcal{E}$ is defined by a set of rules of the form $N \to e$, where e is the expression assigned by $\mathcal{E}$ to non-terminal N. We often write $N \to e$ to mean $e = \mathcal{E}(N)$. In the following, expressions $a \in \Sigma$ and $N \in \mathbb{N}$ will be viewed as special cases of choice expression with $n = 1$. We do not include empty string ε among expressions. The problems caused by it and suggested solutions are discussed in Sect. 2.10.

Non-terminal $N \to e_1 \ldots e_n$ *derives* the string $e_1 \ldots e_n$ of symbols, while $N \to e_1 | \ldots | e_n$ derives one of $e_1, \ldots, e_n$. The derivation is repeated to obtain a string of terminals. This process is represented by syntax tree. The set of all strings derived from $N \in \mathbb{N}$ is called the *language* of N and is denoted by $\mathcal{L}(N)$. Figures 2.1 and 2.2 are examples of grammar G, showing syntax trees of strings derived from the start symbol.

2.3 Recursive Descent

The idea of recursive-descent parsing is that each terminal and each non-terminal is assigned its *parsing procedure*. The procedure either *consumes* some prefix of given input, or fails and leaves the input unchanged. Thus:

- Procedure for terminal a consumes a if the input begins with a. Otherwise it fails.
- Procedure for $N \to e_1 \ldots e_n$ calls the procedures $e_1, \ldots, e_n$, in this order, and fails if any of them fails.

Fig. 2.1 Example of
grammar G and syntax tree
of 'xabbay'

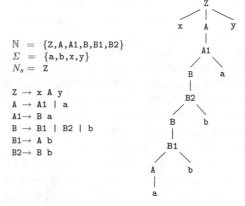

$$\mathbb{N} = \{Z,A,A1,B,B1,B2\}$$
$$\Sigma = \{a,b,x,y\}$$
$$N_s = Z$$

```
Z  → x A y
A  → A1 | a
A1 → B a
B  → B1 | B2 | b
B1 → A b
B2 → B b
```

Fig. 2.2 Example of
grammar G and syntax tree
of 'a*b+a*b'

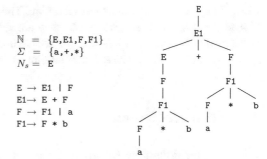

$$\mathbb{N} = \{E,E1,F,F1\}$$
$$\Sigma = \{a,+,*\}$$
$$N_s = E$$

```
E  → E1 | F
E1 → E + F
F  → F1 | a
F1 → F * b
```

- Procedure for $N \to e_1| \dots |e_n$ somehow selects and calls one of the procedures $e_1, \dots, e_n$ that succeeds on the input. It fails if none of them can succeed. (The way of making this selection is discussed in Sect. 2.8.)

An important property of this method of parsing is close connection between the grammar and the parsing procedures. Namely, grammar rules $N \to e_1 \dots e_n$ and $N \to e_1| \dots |e_n$ may be alternatively regarded as concise representations of parsing procedures, with N being the procedure's name and e_i representing a call to procedure e_i. The pattern of successful calls to parsing procedures mimics the derivation process, so successful call to procedure N consumes a prefix belonging to $\mathcal{L}(N)$. The information *how* the consumed string was derived is hidden in the pattern of procedure calls, and recorded in some unspecified way.

We begin our construction by making this recording explicit, and suggest that each successful parsing procedure returns syntax tree of the string it consumed. For this purpose, we add to each procedure its *semantic action*, executed after successful parsing. It is shown below enclosed in braces { }:

$$N \rightarrow e_1 \ldots e_n \; \{\texttt{return}\, [N \lhd [e_1] \ldots [e_n]]\} \tag{2.1}$$

$$N \rightarrow e_1| \ldots |e_n \; \{\texttt{return}\, [N \lhd [e_i]]\} \tag{2.2}$$

We denote here by $[t]$ a syntax tree with top node t, and write $[t \lhd [t_1] \ldots [t_n]]$ to indicate that t has subtrees $[t_1], \ldots, [t_n]$ as children.

Each successfully called procedure e_i returns the syntax tree $[e_i]$. If e_i is for terminal a, it returns syntax tree $[a]$ of height 0. Procedure (2.1) returns a tree with top node N having as children the trees $[e_1], \ldots, [e_n]$ returned by called procedures. Procedure (2.2) returns a tree with top node N having as single child the tree $[e_i]$ returned by the selected procedure.

Parsing starts with invoking (2.1) or (2.2) for the start symbol. It results in recursive calls to subprocedures that eventually return syntax trees for the substrings they consume. At the end, procedure N_s builds and returns the final result.

Assuming that (2.2) makes its choice based only on input ahead, this does not work if the grammar contains left recursion. This is the case for both our examples. If A chooses A1, and B chooses B1, they are called again without any input being consumed in between, and thus bound to indefinitely make the same choices, never returning any result. The situation is similar for E and F.

2.4 Some Definitions

We need some definitions before suggesting how to handle left recursion.

For $N \in \mathbb{N}$ and $e \in \mathbb{N} \cup \Sigma$, define $N \xrightarrow{\text{first}} e$ to mean that procedure N called on some input may call that for e on the same input. We have thus:

- If $N \rightarrow e_1 \ldots e_n$, $N \xrightarrow{\text{first}} e_1$.
- If $N \rightarrow e_1| \ldots | e_n$, $N \xrightarrow{\text{first}} e_i$ for $1 \leq i \leq n$.

Let $\xrightarrow{\text{First}}$ be the transitive closure of $\xrightarrow{\text{first}}$. Non-terminal $N \in \mathbb{N}$ is *(left) recursive* if $N \xrightarrow{\text{First}} N$. The set of all recursive non-terminals of G is denoted by $\mathbb{R}$. All non-terminals in Fig. 2.1 except Z, and all non-terminals in Fig. 2.2 are recursive.

Define relation between recursive N_1, N_2 that holds if $N_1 \xrightarrow{\text{First}} N_2 \xrightarrow{\text{First}} N_1$. This is an equivalence relation that partitions $\mathbb{R}$ into equivalence classes. We call them *recursion classes*. The recursion class of N is denoted by $\mathbb{C}(N)$. All non-terminals in Fig. 2.1 belong to the same class; the grammar of Fig. 2.2 has two recursion classes: {E, E1} and {F, F1}.

In syntax tree, the leftmost path emanating from any node is a chain of nodes connected by $\xrightarrow{\text{first}}$. Suppose N_1 and N_2 belonging to the same recursion class $\mathbb{C}$ appear on the same leftmost path. Any non-terminal N between them must also belong to $\mathbb{C}$, which follows from $N_1 \xrightarrow{\text{First}} N \xrightarrow{\text{First}} N_2 \xrightarrow{\text{First}} N_1$. It means that members of $\mathbb{C}$ appearing on the same leftmost path must form an uninterrupted sequence. We call such sequence a *recursion path* of class $\mathbb{C}$. The only recursion path in Fig. 2.1 is the whole leftmost path from the first A without final a. The syntax tree in Fig. 2.2 has two recursion paths, one starting with E and another with F.

Let $N \to e_1 \ldots e_n$ be on a recursion path. The next item on the leftmost path is e_1, and it must belong to $\mathbb{C}(N)$ to ensure $N \xrightarrow{\text{First}} N$. It follows that the last item on a recursion path must be $N \to e_1 | \ldots | e_n$ where at least one of e_i is not a member of $\mathbb{C}(N)$. Such N is called an *exit* of $\mathbb{C}(N)$, and its alternatives outside $\mathbb{C}(N)$ are the *seeds* of $\mathbb{C}(N)$. In Fig. 2.1, both A and B are exits, and the seeds are a and b. In Fig. 2.2, E and F are exits of their respective classes, and the corresponding seeds are F and a.

A non-terminal that can be the start of a recursion path is called an *entry* of its recursion class. It is one of these:

- An element of $\mathcal{E}(N)$ for non-recursive N.
- One of $e_2, \ldots, e_n$ for recursive $N \to e_1 \ldots e_n$.
- A seed of another recursion class.
- The start symbol.

The recursion class in Fig. 2.1 has A as its entry. The recursion classes in Fig. 2.2 have E and F as their respective entries.

2.5 Recursive Ascent

To handle left recursion, we suggest that parsing procedure for entry E does not follow the pattern given by (2.1) and (2.2) above. It still returns syntax tree $[E]$, but builds this tree in a different way.

As noted in the preceding section, $[E]$ has recursion path starting with the entry node E and ending with the exit node followed by a seed. We start with the seed and ascend the recursion path, adding one node at a time. We reconstruct the side branches when the added node represents sequence expression. The tree being constructed is local to the invocation of procedure E. We call it "the plant", and denote it by $[\pi]$.

Sowing the Seed

Borrowing the way we used to represent parsing procedures, we outline the new entry procedure as:

$$E \to S\{\text{sow } [S]\} \text{ grow} S \; \{\text{return } [\pi]\} . \tag{2.3}$$

Here, S represents call to parsing procedure for the seed S. If successful, it returns syntax tree $[S]$, and semantic action $\text{sow } [S]$ initializes $[\pi]$ with that tree. This is followed by a call to new procedure $\mathrm{grow}S$ that continues parsing and grows the plant towards $[E]$. If it succeeds, the final semantic action returns the reconstructed tree. The whole procedure fails if S or $\mathrm{grow}S$ does.

In general, the recursion class $\mathbb{C}(E)$ may have more than one seed, so E has to choose one that matches the input. To represent this choice, we borrow the syntax of choice expression:

$$E \to (S_1\{\mathrm{sow}\ [S_1]\}\ \mathrm{grow}S_1|\ldots|S_n\{\mathrm{sow}\ [S_n]\}\ \mathrm{grow}S_n)\ \{\mathrm{return}\ [\pi]\}, \quad (2.4)$$

where $S_1, \ldots, S_n$ are all seeds of $\mathbb{C}(E)$.

Growing the Plant

The plant is grown by recursive procedures, each adding one node and then calling another procedure to continue growing. The procedure $\mathrm{grow}R$ that grows plant with top R can be sketched as follows:

$$\mathrm{grow}R \to \mathrm{add}P\ \mathrm{grow}P. \quad (2.5)$$

It adds node P, and continues to grow the plant that has now P on top. The added node P is a predecessor of R in the recursion path, meaning $P \xrightarrow{\text{first}} R$. In general, R may have several predecessors P satisfying $P \xrightarrow{\text{first}} R$, so $\mathrm{grow}R$ must choose one of them that matches the input. Again, we represent this in the way we used for choice expression:

$$\mathrm{grow}R \to \mathrm{add}P_1\ \mathrm{grow}P_1\ |\ \ldots\ |\ \mathrm{add}P_n\ \mathrm{grow}P_n \quad (2.6)$$

where $P_1, \ldots, P_n$ are all members of $\mathbb{C}(E)$ such that $P_i \xrightarrow{\text{first}} R$. We simplify this by introducing new procedure

$$\$P \to \mathrm{add}P\ \mathrm{grow}P, \quad (2.7)$$

so that (2.6) becomes:

$$\mathrm{grow}R \to \$P_1\ |\ \ldots\ |\ \$P_n. \quad (2.8)$$

The growing may stop when the plant reaches E. That means, (2.8) for E must have one more alternative, "do nothing", which we represent by ε:

$$\mathrm{grow}E \to \$P_1\ |\ \ldots\ |\ \$P_n\ |\ \varepsilon. \quad (2.9)$$

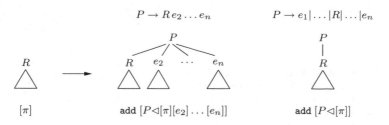

$$P \to R\,e_2 \dots e_n \qquad\qquad P \to e_1 | \dots | R | \dots | e_n$$

Fig. 2.3 add$P$

Adding a Node

Operation add$P$ is illustrated in Fig. 2.3. It adds node P to plant $[R]$.

If $P \to e_1 e_2 \dots e_n$, the added node P will have the present plant as the first child. The other children are obtained by calling $e_2, \dots, e_n$.
Operation add $[P \lhd [\pi][e_2] \dots [e_n]]$ shown in the Figure adds node P with these children to the plant.

If $P \to e_1 | \dots | e_n$, the added node P has only one child, namely the present plant. Operation add $[P \lhd [\pi]]$ shown in the Figure adds such node P to the plant.

Inserting these operations in (2.7), we obtain:

$$\$P \to e_2 \dots e_n \;\{\text{add } [P \lhd [\pi][e_2] \dots [e_n]]\}\; \text{grow}P \quad \text{if } P \to e_1 \dots e_n, \quad (2.10)$$
$$\$P \to \{\text{add } [P \lhd [\pi]]\}\; \text{grow}P \qquad\qquad\qquad \text{if } P \to e_1 | \dots | e_n. \quad (2.11)$$

Multiple Entries

The above assumed that each recursion class has only one entry. This is not true for many grammars; for example, the class of Primary in Java has four entries. Multiple entries can be handled because calls to entry procedures are nested, so one can keep track of which one is currently active. The exit alternative ε in grow$E$ must be applied only if E is the currently active one.

2.6   Examples

Example 1 Applying (2.1), (2.4), (2.8)–(2.9), (2.10)–(2.11) to the grammar of Fig. 2.1, replacing calls for "grow" by their definitions, and omitting unused procedures, we obtain the procedures shown in Fig. 2.4. We apply them to string 'xabbay', showing how they construct the syntax tree appearing in the Figure. Numbers on the left indicate top of the plant after each step. The procedures for

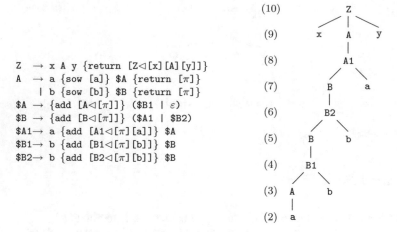

```
Z  → x A y {return [Z◁[x][A][y]]}
A  → a {sow [a]} $A {return [π]}
   | b {sow [b]} $B {return [π]}
$A → {add [A◁[π]]} ($B1 | ε)
$B → {add [B◁[π]]} ($A1 | $B2)
$A1→ a {add [A1◁[π][a]]} $A
$B1→ b {add [B1◁[π][b]]} $B
$B2→ b {add [B2◁[π][b]]} $B
```

Fig. 2.4 Procedures and example of parsing for grammar of Fig. 2.1

choice expressions are assumed to always make correct choice using some oracle; this will be discussed later in Sect. 2.8.

1. The parsing starts with procedure for non-recursive Z that, in the usual way, calls the procedures x, A, and y. After consuming 'x', Z applies A to 'abbay'.
2. Procedure A chooses its first alternative, applies a to 'abbay', leaving 'bbay', and initializes [π] with [a]. Then it calls $A.
3. Procedure $A adds node A on top of [a]. Then it chooses to call $B1.
4. Procedure $B1 applies b to 'bbay', leaving 'bay', and creates node B1 with children [A] and [b]. Then it calls $B.
5. Procedure $B adds node B on top of [B1]. Then it chooses to call $B2.
6. Procedure $B2 applies b to 'bay', leaving 'ay', and creates node B2 with children [B] and [b]. Then it calls $B.
7. Procedure $B adds node B on top of [B2]. Then it chooses to call $A1.
8. Procedure $A1 applies a to 'ay', leaving 'y', and creates node A1 with children [B] and [a]. Then it calls $A.
9. Procedure $A adds node A on top of [A1]. Then it chooses ε, and returns, causing the invoked $A1, $B, $B2, $B, $B1, $A to terminate one after another. Finally, it returns [A] to Z.
10. Procedure Z consumes 'y' and returns the tree [Z◁[x][A][y]].

Example 2 For example of Fig. 2.2, we obtain the procedures shown in Fig. 2.5. We apply them to string 'a*b+a*b', showing how they construct the syntax tree appearing in the Figure. Numbers indicate top of the plant after each step.

1. Procedure E calls parsing procedure for its seed F.
2. Procedure F calls parsing procedure for its seed a, which consumes 'a', leaving '*b+a*b'.
   Then it initializes its plant with [a] and calls $F.

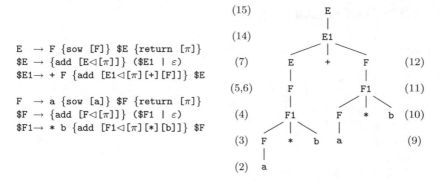

Fig. 2.5 Procedures and example of parsing for grammar of Fig. 2.2

3. Procedure $F adds F on top of [a] and chooses to call $F1.
4. Procedure $F1 calls procedures * and b that consume ′*b′, leaving ′+a*b′.
   Then it adds [F1 ◁ [F][*][b]] on top of [F] and calls $F.
5. Procedure $F adds F on top of [F1] and chooses ε, causing the invoked $F1
   to terminate.
   Then it returns to procedure F.
6. Procedure F returns to E the syntax tree [F] of ′a*b′.
   Procedure E initializes its plant with [F]. What has been the plant of F becomes
   now the seed of E. Procedure E calls then $E.
7. Procedure $E adds E on top of [F] and decides to call $E1.
8. Procedure $E1 calls procedures + and F. The first consumes ′+′, leaving
   ′a*b′.
9. Procedure F calls parsing procedure for its seed a, which consumes ′a′, leaving
   ′*b′. It initializes its own plant with [a]. We have now two plants: that of
   procedure E, for the moment left waiting, and the new plant of F. Procedure F
   calls $F.
10. Procedure $F adds F on top of [a]. Then it chooses to call $F1.
11. Procedure $F1 calls procedures * and b that consume the remaining ′*b′.
    Then it adds [F1 ◁ [F][*][b]] on top of [F] and calls $F.
12. Procedure $F adds F on top of [F1] and chooses ε, causing the invoked $F1
    to terminate.
    Then it returns to procedure F.
13. Procedure F returns to E the syntax tree [F] of ′a*b′.
14. Procedure $E1 adds E1 ◁ [E][+][F] on top of [E]. The plant of F becomes
    now a branch of the plant of E. The procedure calls then $E.
15. Procedure $E adds E on top of $E1 and chooses ε, causing the invoked $E1 to
    terminate. Then it returns to procedure E.
16. Procedure E terminates, returning the constructed tree [E].

2.7 The Dual Grammar

Consider now the procedures in Fig. 2.4 with their semantic actions made invisible:

```
Z   →  x A y
A   →  a $A | b $B
$A  →  $B1 | ε
$B  →  $A1 | $B2
$A1→  a $A
$B1→  b $B
$B2→  b $B
```

This is a new grammar with the set $\{Z, A, \$A, \$B, \$A1, \$B1, \$B2\}$ of non-terminals. Reversing what was said at the beginning of Sect. 2.3, we discover that procedures in Fig. 2.4 are parsing procedures for this grammar. It means that our parsing with recursive ascent is actually performed as recursive descent for a new grammar. This new grammar is in the following referred to as *dual grammar* for the grammar of Fig. 2.1. The procedures differ from (2.1)–(2.2) by having special semantic actions that construct syntax tree according to the original grammar. The dual grammar for grammar of Fig. 2.2 is:

```
E   →  F $E
$E  →  $E1 | ε
$E1→  + F $E
F   →  a $F
$F  →  $F1 | ε
$F1→  * a $F
```

One can see that dual grammar is in general constructed as follows:

- For each entry E create rule $E \rightarrow S_1 \text{ grow}S_1 \mid \ldots \mid S_n \text{ grow}S_n$,
- Replace each recursive $R \rightarrow e_1 e_2 \ldots e_n$ by $\$R \rightarrow e_2 \ldots e_n \text{ grow}R$,
- Replace each recursive $R \rightarrow e_1 \mid \ldots \mid e_n$ by$\$R \rightarrow \text{grow}R$,
- Replace $\text{grow}R$ by $\$P_1 \mid \ldots \mid \$P_m$ if $R \neq E$,
- Replace $\text{grow}E$ by $\$P_1 \mid \ldots \mid \$P_m \mid \varepsilon$,

where

- $S_1, \ldots, S_n$ are all seeds of $\mathbb{C}(E)$,
- $P_1, \ldots, P_m$ are all members of $\mathbb{C}(E)$ such that $P_i \xrightarrow{\text{first}} R$.

The dual grammar is an n-tuple $D = (\mathbb{N}_D, \Sigma, \mathbb{E}_D, \mathcal{E}_D, N_s)$. Its set $\mathbb{N}_D$ consists of:

- The non-recursive members of $\mathbb{N}$;
- The entries to recursion classes;
- The new non-terminals with $-names.

In the following, the set of all non-recursive members of $\mathbb{N}$ is denoted by $\overline{\mathbb{R}}$, and the set of all entries by $\mathbb{R}_E$. They appear as non-terminals in both G and D. The set $\overline{\mathbb{R}} \cup \mathbb{R}_E$ of these common symbols is denoted by $\mathbb{N}_C$. The language $\mathcal{L}_D(N)$ of $N \in \mathbb{N}_D$ is the set of all terminal strings derived from N according to the rules of D.

The two important facts about the dual grammar are:

Proposition 2.1 *The dual grammar is left-recursive only if the original grammar contains a cycle, that is, a non-terminal that derives itself.*

Proof is found in Appendix 1.

Proposition 2.2 $\mathcal{L}_D(N) = \mathcal{L}(N)$ *for all* $N \in \mathbb{N}_C$.

Proof is found in Appendix 2.

2.8   Implementing the Choices

One can see that the choices (2.4) and (2.8)–(2.9) became standard choice expressions in the dual grammar. The dual grammar contains also choice expressions inherited from the original grammar. So far, we did not say anything about how procedures for these expression decide which alternative to choose.

A safe way is to systematically check all choices, backtracking after each failure, until we find the correct one (or find that there is none). This, however, may take exponential time.

There exist shortcuts that give reasonable processing time. The problem is that each is applicable only to a restricted class of grammars. If D is outside this class, the parser will reject some strings of $\mathcal{L}_D(N_s)$.

One shortcut is to look at the next input terminal. This is very fast, and involves no backtracking, but requires that strings derived from different alternatives begin with different terminals. This is known as the LL(1) property. (As a matter of fact, both dual grammars in our examples have this property.)

Another shortcut, not much investigated, is to look at input within the reach of one parsing procedure. It requires that some procedures eventually called from different alternatives never succeed on the same input. This property of the grammar is suggested to be called LL(1p) [8].

Yet another one is limited backtracking, called "fast-back" in [5], page 57, and recently exploited in Parsing Expression Grammar [3]. Here the choice expression tries different alternatives and accepts the first successful one. Backtracking is limited to sequence expressions. Using the technique of "packrat parsing" [2], one can make it work in linear time.

Limited backtracking requires that all choices after first successful one must lead to failure. We say that grammar having this property is fast-back safe. A sufficient condition for fast-back safety is given in [9]. (Note that this property depends on the order of alternatives in choice expressions.)

Proposition 2.2 implies that if grammar D is safe for the chosen method (LL(1), LL(1p), respectively fast-back safe), the recursive-descent parser for D is a correct parser for G.

A subject of further research is how to translate the above properties of grammar D into equivalent properties of grammar G.

2.9　Previous Work

Our method is inspired by one outlined by Hill in [4]. His report describes only the algorithm, but it is clear that it follows from the idea of reconstructing the recursion path. Each entry of recursion class is represented by data structure called the "grower". The grower contains all seeds of its recursion class, and for each seed, the chains of its parents. Parsing procedure for the entry consists of interpreting the grower; it starts with the seeds and follows their parent chains.

The traditional way of eliminating left recursion is to rewrite the grammar so that left recursion is replaced by right recursion. It can be found, for example, in [1], page 176. It consists of replacing

$$A \rightarrow A\ a\ \mid\ b \tag{2.12}$$

by

$$A' \rightarrow b\ A'$$
$$A' \rightarrow a\ A'\ \mid\ \varepsilon\ .$$

Below is an example of this rewriting for the grammar given in [11] as an example of intersecting left-recursion loops:

$$
\begin{array}{ll}
& E \rightarrow\ F\ n\ \mid\ n \\
E \rightarrow\ F\ n\ \mid\ n & F \rightarrow\ (n\ +\ x\ \mid\ G\ -)\ F' \\
F \rightarrow\ E\ +\ x\ \mid\ G\ - & F' \rightarrow\ (n\ +\ x)\ F'\ \mid\ \varepsilon \\
G \rightarrow\ H\ m\ \mid\ E & G \rightarrow\ (H\ m\ \mid\ n\ \mid\ (n\ +\ x)\ F'\ n)\ G' \\
H \rightarrow\ G\ y & G' \rightarrow\ (-\ F'\ n)\ G'\ \mid\ \varepsilon \\
& H \rightarrow\ (n\ G'\ y\ \mid\ (n\ +\ x)\ F'\ n\ G'\ y)\ H' \\
& H' \rightarrow\ (m\ G'\ y)\ H'\ \mid\ \varepsilon
\end{array}
$$

The process is cumbersome and produces large results. This can be compared with our dual grammar, which is obtained by a rather straightforward rewriting of syntax rules:

$$
\begin{array}{l}
E\ \rightarrow\ n\ \$E \\
\$E\ \rightarrow\ +\ x\ \$F\ \mid\ \$G\ \mid\ \varepsilon \\
\$F\ \rightarrow\ n\ \$E \\
\$G\ \rightarrow\ -\ \$F\ \mid\ \$H \\
\$H\ \rightarrow\ y\ m\ \$G
\end{array}
$$

The main problem with the traditional rewrite is that it completely loses the original intent of the grammar. The same can be said about our dual grammar, but the difference is that our recursive descent for dual grammar produces correct syntax tree according to the original grammar.

In [13], Warth and al. suggested how to handle left recursion in Parsing Expression Grammar (PEG) by modifying the packrat parser. The principle of packrat parsing is that the result of each parsing procedure is saved in a "memoization table". The table is consulted before each new call, and if a result is found for the same position in input, it is used directly instead of calling the procedure.

The idea can be explained on the example grammar (2.12). Before the first call to A, we save "failed" as the result of A at the current position. Procedure A starts with the alternative Aa which requires calling A and then a. But the result of calling A is already in the table; as it is "failed", the alternative Aa fails, and A calls b. If it succeeds, we save 'b' as the new result of A.

We backtrack and call A again. Alternative Aa obtains now 'b' as the result of A and calls a. If it succeeds, we save 'ba' as the new result of A. This is referred to as "growing the seed".

We repeat this as long as a in Aa succeeds. When it fails, we stop the repetition and are left with the result 'ba...a' of A in the table.

In [12], Tratt indicated that this method does not require packrat parser, only a table for results of left-recursive procedures. Also, that left-recursive non-terminals can be detected by analyzing the grammar.

Medeiros et al. [7] introduced the notion of "bounded left recursion". The idea can again be explained on the example grammar (2.12). Procedure A is required to try alternative Aa exactly n times before choosing b. The string thus consumed is saved as A^n. We call procedure A repeatedly with increasing values of the "bound" n, obtaining A^0, A^1, A^2, etc..

Left part of Fig. 2.6 illustrates how A^2 is obtained for input 'baax'. The arrows represent successful calls. One can see from the Figure how this is obtained from saved $A^1 = $ 'ba', and this from $A^0 = $ 'b'.

Right part of Fig. 2.6 shows what happens for $n = 3$. Dotted arrows represent failing calls. The a in the highest Aa encounters 'x' and fails, and so does Aa. Procedure A, receiving failure from Aa, consumes 'b', giving $A^3 = $ 'b'. This is the signal to stop repetition and return the saved A^2.

The methods from [7, 12, 13] use memoization table; they repeatedly evaluate the left-recursive procedure in the process of growing the seed or incrementing the bound. None of these is the case with our approach.

All methods [4, 7, 12, 13] must handle situations more complex than the elementary example (2.12), like indirect and nested left recursion. They add a substantial amount of overhead to the mechanism of procedure call. This is not the case with our method, where all complexity is delegated to the task of constructing the dual grammar. The only overhead are the semantic actions of constructing the syntax tree, but they must also be invisibly included in the above methods to record the pattern of procedure calls.

Fig. 2.6 Example of bounded left recursion applied to 'baax'

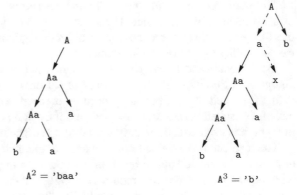

In principle, we have a recursive-descent parser with simple calls to parsing procedures. If so desired, it can use limited backtracking implemented with packrat technology.

In addition, we offer a way to see if the chosen method of handling choice expressions results in correct parser. None of the cited solutions gives any hint on how to do it.

2.10 Unsolved Problems

We did not include empty word ε in our grammar. (It appeared in the dual grammar, but only with a special function.) The result of adding ε is that some expressions may derive empty string. These expressions are referred to as *nullable*, and can be identified as such by analyzing the grammar.

Nullable e_1 in $N \to e_1 \ldots e_n$ and nullable e_i in $N \to e_1| \ldots |e_n$ extend the definition of $\xrightarrow{\text{first}}$ beyond the two rules given in Sect. 2.4. As the result, non-terminal may be recursive or not depending on input.

An example is $A \to (a \mid \varepsilon) Ab \mid c$ where A is left-recursive only if input does not start with a. It is known as "hidden" left recursion. This specific case can be solved by redefining the grammar as

$$A \to (Ab|c) \mid (aAb|c),$$

but we do not provide a general solution. One can always allow ε under condition that it does not affect the set of recursive non-terminals, which can be checked by examining the grammar.

Nullable seed and nullable $e_2 \ldots e_n$ in recursive $N \to e_1 e_2 \ldots e_n$ invalidate the proof of Proposition 2.1. That means the dual grammar can be left-recursive even if G does not have a cycle. It requires a specific check for D being left-recursive.

Another kind of unsolved problem is E → E+E|n, which results in a right-recursive parse for E. This is so because the second E invoked by (2.10) gobbles up all input before the ascent can continue.

The problem with E → E+E|n was signalled by Tratt in [12]. In fact, the right-recursive result is correct because the grammar is ambiguous. Tratt explains why the left-recursive parse is desirable, and offers a solution for direct left recursion.

A solution proposed in [7] extends the grammar by assigning "priorities" to different occurrences of E. This opens new possibilities: a simple way of defining operator precedence. It is further exploited in [6].

Acknowledgements This version is the result of comments from two anonymous referees and comments to presentation at CS&P'2021 workshop [10].

2.11 Appendix 1: Proof of Proposition 2.1

For $N \in \mathbb{N}_D$ and $e \in \mathbb{N}_D \cup \Sigma$, define $N \xrightarrow{\text{firstD}} e$ to mean that parsing procedure N may call parsing procedure e on the same input:

(a) For $N \in \overline{\mathbb{R}}$, $\xrightarrow{\text{firstD}}$ is the same as $\xrightarrow{\text{first}}$.

(b) For $N \in \mathbb{R}_E$, $N \xrightarrow{\text{firstD}} S$ for each seed S of $\mathbb{C}(N)$.

(c) For $\$R \to e_2 \ldots e_n \text{grow} R$, $\$R \xrightarrow{\text{firstD}} e_2$.

(d) For $\$R \to \text{grow} R$, $\$R \xrightarrow{\text{firstD}} \$P_i$ for each $P_i \in \mathbb{C}(R)$ such that $P_i \xrightarrow{\text{first}} R$.

Suppose D is left-recursive, that is, there exist $N_1, N_2, \ldots, N_k \in \mathbb{N}_D$ where $N_1 \xrightarrow{\text{firstD}} \ldots N_2 \xrightarrow{\text{firstD}} \ldots \xrightarrow{\text{firstD}} N_k$ and $N_k = N_1$. We start by showing that none of them can be in $\mathbb{N}_C$.

Assume that $N_1 \in \mathbb{N}_C$. That means N_1 is either in $\overline{\mathbb{R}}$ or in $\mathbb{R}_E$.

Suppose $N_1 \in \overline{\mathbb{R}}$. Because N_1 is the same in both grammars, we have $\mathcal{E}_D(N_1) = \mathcal{E}(N_1)$, so N_2 is also in $\mathbb{N}$, and thus in $\mathbb{N}_C$. From $N_1 \xrightarrow{\text{firstD}} N_2$ follows $N_1 \xrightarrow{\text{first}} N_2$.

Suppose $N_1 \in \mathbb{R}_E$. According to (b), N_2 is a seed of $\mathbb{C}(N_1)$, which is either non-recursive or an entry, and thus is in $\mathbb{N}_C$. For a seed N_2 of $\mathbb{C}(N_1)$ holds $N_1 \xrightarrow{\text{First}} N_2$.

The above can be repeated with $N_2, \ldots, N_{k-1}$ to check that $N_i \in \mathbb{N}_C$ for all $1 \le i \le k$ and $N_i \xrightarrow{\text{First}} N_{i+1}$ for $1 \le i < k$. As $N_1 = N_k$, the latter means that none of N_i belongs to $\overline{\mathbb{R}}$, so they must all be in $\mathbb{R}_E$. Moreover, they are all in the same recursion class, $\mathbb{C}(N_1)$. According to (b), N_2 is a seed of $\mathbb{C}(N_1)$, that cannot be a member of $\mathbb{C}(N_1)$. This is a contradiction, so $N_1 \notin \mathbb{N}_C$, and this holds for the remaining $N_2, \ldots, N_{k-1}$.

It follows that each N_i is a \$-non-terminal of D. It cannot be that listed under (c) because then N_{i+1} is e_2 which belongs to $\mathbb{N}_C$.

Hence, each N_i must have the form listed under (d). It follows that if $N_i = \$R_i$ for some R_i, N_{i+1} must be $\$R_{i+1}$ such that $R_{i+1} \xrightarrow{\text{first}} R_i$. Thus, exists in grammar

G a sequence of non-terminals $R_k \xrightarrow{\text{first}} \ldots \xrightarrow{\text{first}} R_2 \xrightarrow{\text{first}} R_1$ with $R_k = R_1$. Each of them has choice statement as $\mathcal{E}(R_i)$ and can derive itself in $k - 1$ steps.   □

2.12 Appendix 2: Proof of Proposition 2.2

We show that each string derived from $N \in \mathbb{N}_C \cup \Sigma$ according to grammar G can be derived according to grammar D and vice-versa. We say that derivation has height h to mean that its syntax tree has height h. The proof is by induction on that height.

Induction base is the same in both directions: derivation of height 0. We consider terminal to be derived from itself with syntax tree of height 0. The G- and D-derivations are identical.

The following two lemmas provide induction step for the two derivations.

Lemma 2.1 *Assume that each string having G-derivation of height $h \geq 0$ from $N \in \mathbb{N}_C \cup \Sigma$ has a D-derivation from N. Then the same holds for each string with G-derivation of height $h + 1$.*

Proof Take any string w having G-derivation from $N \in \mathbb{N}_C$ of height $h + 1$.

(Case 1) $N \in \overline{\mathbb{R}}$.

If $N \to e_1 \ldots e_m$, all e_j are in $\mathbb{N}_C$ and $w = w_1 \ldots w_m$ where each w_j is G-derived from e_j by derivation of height h or less. We D-derive $e_1 \ldots e_m$ from N, and then, according to induction hypothesis, D-derive $w = w_1 \ldots w_m$ from $e_1 \ldots e_m$.

If $N \to e_1 | \ldots | e_m$, w is G derived from one of e_j by derivation of height h or less. We D-derive e_j from N, and then, according to induction hypothesis, D-derive w from e_j.

(Case 2) $N \in \mathbb{R}_E$.

The G-derivation of w is shown in the left part of Fig. 2.7. One can see that $w = w_0 \ldots w_n$ where w_0 is G-derived from a seed and for $1 \leq i \leq$, w_i is either G-derived from $e_2 \ldots e_m$ (if $R_i \to e_1 \ldots e_m$) or is empty word ε (if $R_i \to e_1 | \ldots | e_m$). All these G-derivations are from members of $\mathbb{N}_C$ and have height h or less. Thus, by induction hypothesis, exist corresponding D-derivations of $w_0, \ldots, w_n$. The D-derivation of w using these results is shown as the right part of Fig. 2.7.   □

Lemma 2.2 *Assume that each string having D-derivation of height $h \geq 0$ from $N \in \mathbb{N}_C \cup \Sigma$ has a G-derivation from N. Then the same holds for each string with D-derivation of height $h + 1$.*

Proof (The proof is a mirror image of the proof of Lemma 2.1, but we spell it out.) Take any string w having D-derivation from $N \in \mathbb{N}_C$ of height $h + 1$.

(Case 1) $N \in \overline{\mathbb{R}}$.

If $N \to e_1 \ldots e_m$, all e_j are in $\mathbb{N}_C$ and $w = w_1 \ldots w_m$ where each w_j is D-derived from e_j by derivation of height h or less. We G-derive $e_1 \ldots e_m$ from N, and then, according to induction hypothesis, G-derive $w = w_1 \ldots w_m$ from $e_1 \ldots e_m$.

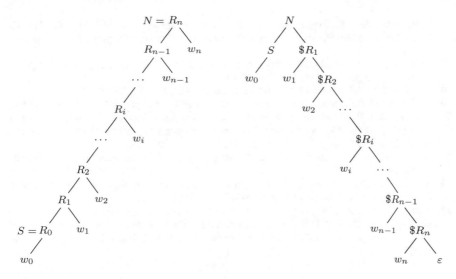

Fig. 2.7 G-derivation and D-derivation from E. w_0 is derived from the seed. Otherwise w_i is derived from $e_2, \ldots, e_m$ for sequence R_i or ε for choice R_i

If $N \to e_1 | \ldots | e_m$, w is D-derived from one of e_j by derivation of height h or less. We G-derive e_j from N, and then, according to induction hypothesis, G-derive w from e_j.

(Case 2) $N \in \mathbb{R}_E$.

The D-derivation of w is shown in the right part of Fig. 2.7. One can see that $w = w_0 \ldots w_n$ where w_0 is D-derived from a seed and for $1 \le i \le$, w_i is either D-derived from $e_2 \ldots e_m$ (if $R_i \to e_1 \ldots e_m$) or is empty word ε (if $R_i \to e_1 | \ldots | e_m$). All these D-derivations are from members of $\mathbb{N}_C$ and have height h or less. Thus, by induction hypothesis, exist corresponding G-derivations of $w_0, \ldots, w_n$. The G-derivation of w using these results is shown as the left part of Fig. 2.7. □

References

1. Aho, A.V., Sethi, R., Ullman, J.D.: Compilers, Principles, Techniques, and Tools. Addison-Wesley (1987)
2. Ford, B.: Packrat parsing: simple, powerful, lazy, linear time, functional pearl. In: Wand, M., Jones, S.L.P., (eds.) Proceedings of the Seventh ACM SIGPLAN International Conference on Functional Programming (ICFP '02), Pittsburgh, Pennsylvania, USA, October 4–6, 2002, pp. 36–47. ACM (2002)
3. Ford, B.: Parsing expression grammars: A recognition-based syntactic foundation. In: Jones, N.D., Leroy, X., (eds.) Proceedings of the 31st ACM SIGPLAN-SIGACT Symposium on Principles of Programming Languages, POPL 2004, pp. 111–122. ACM, Venice, Italy (2004)
4. Hill, O.: Support for Left-Recursive PEGs (2010). https://github.com/orlandohill/peg-left-recursion
5. Hopgood, F.R.A.: Compiling Techniques. MacDonald/Elsevier (1969)

6. Laurent, N., Mens, K.: Parsing expression grammars made practical. In: SLE 2015: Proceedings of the 2015 ACM SIGPLAN International Conference on Software Language Engineering, pp. 167–172 (2015)
7. Medeiros, S., Mascarenhas, F., Ierusalimschy, R.: Left recursion in parsing expression grammars. Sci. Comput. Program. **96**, 177–190 (2014)
8. Redziejowski, R.R.: From EBNF to PEG. Fund. Inf. **128**, 177–191 (2013)
9. Redziejowski, R.R.: More about converting BNF to PEG. Fund. Inf. **133**(2–3), 177–191 (2014)
10. Redziejowski, R.R.: Left recursion by recursive ascent. In: Schlingloff, H., Vogel, T., (eds.), Proceedings of the 29th International Workshop on Concurrency, Specification and Programming (CS & P 2021), pp. 72–82 (2021). http://ceur-ws.org/Vol-2951
11. Sigaud, P.: Left recursion (2019). https://github.com/PhilippeSigaud/Pegged/wiki/Left-Recursion
12. Tratt, L.: Direct left-recursive parsing expression grammars. Tech. Rep. EIS-10-01, School of Engineering and Information Sciences, Middlesex University (2010)
13. Warth, A., Douglass, J.R., Millstein, T.D.: Packrat parsers can support left recursion. In: Proceedings of the 2008 ACM SIGPLAN Symposium on Partial Evaluation and Semantics-based Program Manipulation, PEPM 2008, San Francisco, California, USA, January 7-8, pp. 103–110 (2008)

Chapter 3
An Example of Computation in Interactive Granular Computing

Soma Dutta

Abstract This paper is an attempt to clarify the role of Interactive Granular Computing (IGrC) as a computation model through a real life example. The model emphasizes on the consideration that a real cognition about a real physical complex phenomenon and decision making in such a complex system cannot be formalized only being in the language of mathematics.

3.1 Introduction

This paper is an extension of the paper [7], in which the model of computation in has been explained in the context of a blind person. A blind person in order to move forward without stumbling over the obstacles, needs to explore and learn the surrounding environment through interactions with the environment. In continuation to [7], this paper explains in more detail different components of the IGrC model with respect to the mentioned example and explains how it has the potential of handling the grounding problem by bridging a connection between abstract mathematical modeling and real physical semantics.

In some of the previous papers [8, 9, 19], including [7], the arguments in favour of a model for computing and reasoning, which is not purely mathematical and isolated from its real physical semantics and which has the possibility to learn from the real physical environment through real physical interactions, are discussed. The need for such a model in building an intelligent system is also supported by several researchers [1–6, 10–13, 16, 17]. One of these work [17] clearly articulates about what is missing in the theories of symbolic manipulations, which usually prevails in practice.

> [...] *the question of the bridge between physical and symbolic descriptions determines in a fundamental way how one views the problems of cognition. A primary question here is, Exactly what kind of function is carried out by that part of the organism which is variously called the sensor, the transducer, ..., unlike the symbolic processes that characterize*

S. Dutta (✉)
University of Warmia and Mazury in Olsztyn, Sloneczna 54, 10-710 Olsztyn, Poland
e-mail: soma.dutta@matman.uwm.edu.pl

© The Author(s), under exclusive license to Springer Nature Switzerland AG 2023
B. Schlingloff et al. (eds.), *Concurrency, Specification and Programming, Studies in Computational Intelligence 1091*, https://doi.org/10.1007/978-3-031-26651-5_3

cognition, the function performed by the transducer cannot be described ... by a purely
symbol-manipulation function. ... the transducer fundamentally is a physical process

Till now, in different works (see, *e.g.*, [8, 9, 14, 18, 19]), we tried to introduce what do we mean by *Interactive Granular Computing* (IGrC) and how it is different from the other existing theories from the perspective of modeling computations in a complex system. In a very brief description, *Interactive* symbolizes interaction between the abstract world and the real physical world, and *Granular Computing* symbolizes computation over imperfect, partial, granulated information abstracted about the real physical world.

The notion of *complex granule* (c-granule) is the basic building block of IGrC. In the real world, a real physical fragment and the physical interactions within it together creates a complex dynamic system. An intelligent agent, lying within the mentioned fragment (or outside of it), may need to gather information about the fragment, initiate real physical actions over it, and take decisions about next plan of actions based on that. Here both the agent and the fragment of the real physical world can be considered as (). The former may have access and control over its own information base and is called ; while the respective information base of the latter may not be accessed by it. But in the process of any computation over this fragment of the real physical world, both of these c-granules may need to interact, and accordingly, in order to plan further actions or decisions, the information base of the ic-granule involved in the process needs to be updated. Such a process of computation represents a complex phenomenon, which can involve both physical and informational objects, and their interactions in order to learn and perceive the state of the environment where the computation is running.

In this paper, our main focus is to present a real life example, through which different aspects of computation over a c-granule can be visulalisd. In this regard, Sect. 3.2 presents an overview of c-granule and basic relevant notions of IGrC model, and Sect. 3.3 presents an exampe of a real life computation from the perspective of IGrC. Section 3.4 presents the concluding remarks.

3.2 Basics of c-granule, ic-granule, and Control

Here, we attempt to clarify the intuition behind the notion c-granule and other associated terms. A c-granule bases on a collection of physical objects and a set of informational objects describing some information about those physical objects. The collection of physical objects determines the, which may vary over time. For some c-granules, the information regarding some of the objects from its scope is attached to it as an information layer, and those c-granules are known as ic-granules; otherwise it is simply called a c-granule.

The information related to a c-granule is used to control the behaviour and interaction of the c-granule with other c-granules. This control mechanism is known as the *control* of a c-granule, and the may itself be located at the same c-granule or at some other c-granule. A c-granule containing its own control is known as a c-granule with control. A c-granule having its control located at some other c-granule corresponds to such complex phenomenon where the control of an event is performed remotely. A formal account of the notion of control of a c-granule is presented in Sect. 3.3.1.

A c-granule is divided into three parts, namely, soft_suit, link_suit, and hard_suit, and they correspond to three collections of physical objects from the scope of the concerned c-granule. The collection of physical objects, which are directly accessible and/or about which already some information is obtained by the control, pertains to the soft_suit. The objects, which are in the scope of the c-granule but not yet accessed or are not in the direct reach at that point of time of the c-granule, belong to the hard_suit. The link_suit, on the other hand, consists of a network or chain of objects that creates a communication channel between the soft_suit and the hard_suit. In Fig. 3.1, the structure of an ic-granule is presented in the context of an example.

The contains different forms of information, such as specifications of the already perceived properties of the objects from its soft_suit, specifications of the windows describing where and how some specific part from the scope of the ic-granule can be accessed, measured or reached etc. Based on the purposes and types of the information specifications of an ic-granule they can be classified in different categories, such as ic-granule representing perception of the objects from the scope (perception based ic-granule), ic-granule representing domain knowledge (knowledge based ic-granule), ic-granule representing plan of actions (planner ic-granule), ic-granule representing plan into implementational level language (implementational ic-granule) etc. Here, to be emphasized that the ic-granule, responsible for implementation of a plan of actions, serves the task of connecting between the abstract and the real worlds.

3.2.1 An Overview of Computation Over c-granule

A computation process over a c-granule is described by a network of ic-granules having their scope within the scope of the concerned c-granule. The informational layers of all these ic-granules constitute the domain knowledge of the *control* of the c-granule. So, the computation running over a c-granule is basically led by the control of the c-granule. That is, why in other words we may also call the control of the c-granule, over which the computation process is running, as the control of the computation process. Apart from this informational content the control is also endowed with a reasoning mechanism.

The information layer of each ic-granule is clustered based on the information relevant to different sub-scopes of the whole scope of the ic-granule. For example, in the ic-granule representing the domain knowledge, there can be different sub-clusters in the informational layer corresponding to different aspects of the domain knowledge. That is, an ic-granule may contain several other ic-granules inside its scope.

The reasoning mechanism of the control of a c-granule is responsible for aggregating, deleting, or generating new information from the existing information, generating action plans, converting the plan of actions to an implementational level language, and updating new information that is perceived after implementaion of actions. Thus new information layers are generated over time based on

(i) initial (partial) perception and domain knowledge of the concerned fragment of the environment,
(ii) initiation of the interactions through already accessible objects in order to access information about the not directly reachable objects,
(iii) perception of the physical world after interactions, and
(iv) verification of the perceived properties of the newly obtained configuration with the expected specifications of the target environment.

In particular, the control works on the basis of the gathered information in the informational layer of all the ic-granules within the scope of the concerned c-granule. However, in a single step it may access only a part of the information layer where the information about the currently perceived situation is represented in some language. If this information is not sufficient for selection of the relevant rules to transform the current configuration of the ic-granules into the required one, then the control aims at extending this information. For learning new information, it may use the already stored knowledge base and/or decide to perform necessary measurements or actions in the relevant fragments of the physical world. These relevant fragments of the physical world are also indicated by some spatio-temporal windows, with the help of which the control can point to the respective fragments of the world, stored in the information layer of the control.

Combining the above mentioned features of computing under IGrC, one can notice that in one hand it uses the informational layers of the perception based ic-granule, the knowledge based ic-granule and the planner ic-granule to generate a specification of the desired plan of actions; on the other hand, through the control mechanism it translates the specification of the abstract plan of actions in the language of the actuators, and incorporates a special feature, which considers that going beyond the mathematical specifications of actions the real physical actions will be embedded through the real physical actuators. Here comes the part related to real physical phenomenon which is out of the realm of mathematics. To be noted that the specifications of the plan of actions, both the abstract specifications and specifications in the language of the actuators, are part of the IGrC model; however, the real physical actions, which are supposed to be embedded on some real physical objects following the specifications of the actions, are not part of the mathematical manifold. These actions are out of the realm of the IGrC model.

Thus the process of computation in IGrC allows a time lapse between the prescribed plan to be transferred from the abstract module to the implememtational module, and some real physical interactions take place after that transformation of the plan. At the end of this time interval, a new cycle of the control starts by perceiving the changes occurred in the environment by the interactions initiated in the concerned interval of time. After each such real physical interaction the newly perceived properties of the real physical environment are verified with the specification of the expected outcome available in the information layer of the concerned ic-granule. Based on how satisfactorily the perceived properties match to the specification of the expected properties, a new plan of actions is generated. All these changes, such as generation of new planner ic-granule, updating new information in the information layer of an ic-granule, happening during the process of computation, indicate a change in the configuration[1] of the ic-granules over the passage of the time.

The above description of the computation over a c-granule clarifies how the structure of the initial network of ic-granules changes over time by changing either the informational layer of some of its ic-granules or by having a new ic-granule obtained from a previous ic-granule by initiation of interactions in the scope of the latter. In Sect. 3.3 we shall present a detailed exemplification of how these different ic-granules such as perception based ic-granule, knowledge based ic-granule, planner ic-granule, implementational ic-granule etc are composed with three collections of physical objects and the respective informational layers, as well as how a computation running over such a network bring changes in their configurations over time.

3.3 Computation Over c-granules: A Real Life Example

In [15], a very nice exemplary context of computation has been presented to explain how a model of computation for simulating real life phenomenon should look like.

> *perceiving is a way of acting.* [. . .] *Think of a blind person tap-tapping his or her way around a cluttered space, perceiving that space by touch, not all at once, but through time, by skillful probing and movement. This is or ought to be, our paradigm of what perceiving is.*

The above cited example seems very appropriate to reflect the concern that we attempt to address through the notion of . According to the cited example, learning about the surrounding environment depends on a process of exploration through continuous process of perception and deriving the next move based on that. Thus, it endorses that in order to obtain information about the environment, an intelligent agent needs to perform a continuous process of real physical interactions through some actuators; in the context of this example the actuator can be considered as the stick of the blind person.

So, casting the above mentioned context in the framework of computation over c-granules, we can notice that the perception based ic-granule pertains to the blind person, the stick and the surrounding environment. At time t_0 of starting the movement some information about the already perceived objects from the environment may be attached to this particular ic-granule. To gather more information about the

---

[1] The configuration of an ic-granule means the collection of physical objects in the scope of the ic-granule as well as the information layer attached to it. Thus change in the configuration of an ic-granule means change in either of its parts.

environment some real physical interactions, through the stick, may be required. The person may have some knowledge base set into its brain. So, the knowledge based ic-granule can be considered as the brain of the blind person where it can have directly accessible memory cells as well as some dormant memory cells, which need to be activated by the process of thinking and analyzing. Based on the information of the environment, gathered from the sensors of the stick and attached to perception based ic-granule, as well as the information attached to the knowledge based ic-granule, the agent may select some specific actions. The specification of the plan of actions is attached to the planner ic-granule, whose scope is again some part of the persons' brain-cells. In case of lack of choice of any relevant rule the agaent may continue the probing further. That means, the planner ic-granule may generate a further spec-ification for measuring and perceiving the environment. The abstract specification generated by the planner ic-granule gets transferred to the where it is stored in a specifying the actions that to be performed through the actuators. Intuitively, this indicates the transformation of the plan, from the blind person's brain cells, to the language of the action specification of the stick. For instance, if the abstract plan is *to move forward* then its respective specification in the language of the stick may be *moving the end of the stick along a straight line perpendicular to the horizontal axis passing through the tip of the stick*. The next step considers that this action specification for the stick must be embedded through the stick to the specified real physical fragment of the concerned perception based ic-granule. This brings in the feature of real physical interactions in the model.

So, we see that apart from an informational base, the continuous contact and interactions with the physical reality play a crucial role in the process of deriving next plan of action. Hence, in order to build an intelligent agent, simulating decision making ability like an human, all the components such as, process of perception, knowledge base, real physical actions, etc are to be accommodated together in the model of such an agent.

The above attitude towards building an intelligent agent certainly matches to our perspective of computation over c-granule. So, as an example of a computation over a c-granule, we choose this particular example from the above quotation. Here instead of a blind person, as the scope of the c-granule, we can simply think of a robot with a stick placed in a real physical environment. Moreover, the robot can be considered a c-granule with control.

Specifically, the robot and the top part of the stick are directly accessible by the control of the c-granule and hence belongs to the soft_suit. The part of the stick, which is distant from the direct touch of the robot, belongs to the link_suit as it creates a link to the not directly accessible objects, such as holes or stones lying in the surrounding environment, that is to the hard_suit. The goal of the computation is to have a successful, safe movement of the robot by deriving information about the unseen objects based on the already available knowledge and the perceived informa-tion about the directly accessible objects. The whole computation process, leading to the goal, is conducted by the control of the mentioned c-granule.

All the information layers of the ic-subgranules of the given c-granule together with the reasoning mechanism of the control makes the computation to move dynamically from one layer or stage to another over time. Moving from one layer to another basically indicates the progress of the computation, and this is reflected by the way the control transforms a finite collection or network of ic-granules to the new ones. These networks of ic-granules may contain perception based ic-granule, knowledge based ic-granule, planner ic-granule, and implementational ic-granule. With the progress of the computation each of those ic-granules may have changes in their physical configurations or in the informational layers. The change in the physical configuration of an ic-granule happens when through the implementational ic-granule real physical actions are embedded on a real fragment of the world. The realization of those actions are not formalized in the framework of the IGrC model, but it incorporates the possibility of perceiving the new configuration of the environment after the initiation of the actions, and thus reasons based on the newly gathered information. This in turn, brings a change in the information layer of the concerned ic-granule. Thus from one time point to the next new network of ic-granules are generated.

Below a step-by-step process of the computation over a c-granule, in the context of movement of the robot through probing and perceiving the surrounding environment with the help of the stick, is described.

3.3.1 Control of a c-granule: A Formal Account

Before, presenting the case specific description of the whole process of computation over a c-granule, led by its control, let us first present a formal account of the notion of control of a c-granule. Intuitively, the control of a c-granule also have an informational aspect and a physical aspect. That is, some components in it can manipulate and generate information based on the current perception and the knowledge base, and some components in it can translate those information in the language of the actuators and create an interface for embedding those actions on some real physical objects. This interface is responsible for real physical interactions. After each real physical interactions a new cycle of control starts with perceiving and updating information about the environment. Such cycles of the computation conducted by the control can be marked by a successive points of time.

Definition 3.1 The control of a c-granule g at a time point t, denoted as $Cont_t(g)$, is represented by a tuple $(Nt_t, Lan_{upp}, Lan_{low}, \mathcal{R}_{inf}, \mathcal{R}_{imp}, Mod_{imp})$ where the components are defined as follows.

(i) Nt_t is a network of ic-granules, may or may not containing g.[2]
(ii) Lan_{upp} is an for representing the specifications of actions.
(iii) Lan_{low} is a lower level language representing each specification from Lan_{upp} into a language of actuators, or language of implementation.

---

[2] The context when $Cont_t(g)$ refers to a network of ic-granules not containing the c-granule g, is analogous to the situation when a real physical complex phenomenon is controlled by another remote physical phenomenon.

(iv) $\mathcal{R}_{inf} \subseteq Inf(Nt_t) \times Lan_{upp}$ where $Inf(Nt_t)$ denotes the union of all the
        information layers of the ic-granules in Nt_t. Given a set of information from
        the information layers of the ic-granules from Nt_t, $\mathcal{R}_{inf}$ specifies an action
        plan in Lan_{upp}.

(v) $\mathcal{R}_{imp} \subseteq Lan_{upp} \times Lan_{low}$ contains the relational pairs indicating the transla-
        tion of the abstract specification of an action into the language of the imple-
        mentation.

(vi) Mod_{imp} represents a module which takes the instructions generated by $\mathcal{R}_{imp}$
        and then implements those instructions by embedding them on the relevant
        real physical objects.

So, we can observe that $Cont_t(g)$ is endowed with a network of ic-granules, an
upper level language describing the abstract specifications of the process of computation, a lower level language describing the actions for implementing those abstract
specifications through real physical actuators, two relations $\mathcal{R}_{inf}$ and $\mathcal{R}_{imp}$, and a
module, namely Mod_{imp}, responsible for the real physical actions; $\mathcal{R}_{inf}$ determines
the abstract specifications of the plan of actions based on the available information
regarding perception and the domain knowledge, and given the abstract specifications of the plan of actions $\mathcal{R}_{imp}$ determines the plan of action in the language of
implementation. Here to be remembered, that though $\mathcal{R}_{imp}$ generates a specification
for the plan of actions in an implementational level language, embedding that plan
of actions through a real physical object remains outside the realm of mathematical
formalization. So, the structure and functionalities of the implementation module
Mod_{imp} may be considered more like a technically equipped module than a mathematical construct. We only assume that some real physical interactions must happen
by embedding the plan of actions over the objects lying in the scope of the c-granule,
and the control of g only can gather new information by perceiving the environment
after the completion of those real physical interactions. Moreover, the component
Nt_t also indicates a network of ic-granules, which are not considered as the so-called
mathematical objects. So, this clarifies how the control is constructed out of an and
a, which through a continuous process of real physical interactions connect among
abstract informations about the physical reality and its actual perceived states.

3.3.2   Computation in the Context of the Robot with a Stick

Now we focus on explaining and relating the notions such as network of ic-granules,
computation over ic-granules conducted by the control etc in the context of the
example of moving of a robot with a stick in a partially known environment. Here
for the sake of ease of understanding we prefer to present a simplified example of
control and its role, which in the case of real life scenario may have more complex
structural and functional aspects.

Computation at time t_0: Let us now present all the components of the computation
    present at the starting time, say t_0 of the computation in the context of the robot

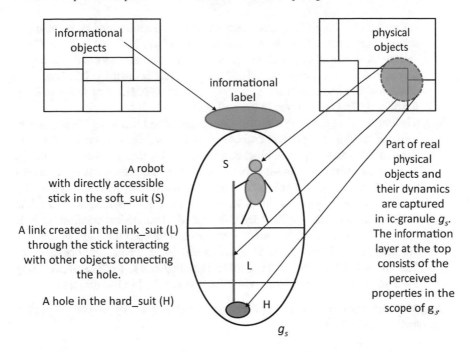

Fig. 3.1 The structure of an ic-granule, denoted by g_s; the scope of it contains a robot with a stick and the surrounding objects in the concerned real physical fragment. The symbols S, L, and H respectively denote the soft_suit, link_suit, and hard_suit of g_s. The information layer attached to it contains information regarding already perceived objects and some general rules relevant for the environment

exploring the environment through its stick. In particular, below we first introduce the concerned ic-granules at time t_0.

1. Let g_{perc} be an ic-granule having in its scope a robot with a stick and the objects lying in the surrounding of the robot. At t_0, the beginning of the control's cycle, the information layer of g_{perc} is labelled with the perceived information of the directly accessible objects from its soft_suit. As mentioned before, here the directly accessible objects can be the robot and the part of the stick directly in contact with it (see Fig. 3.1). The ic-granule g_{perc} represents a perception based ic-granule and its information layer contains, in particular, information about the perception of its environment at time t_0, domain knowledge related to the general perception of the environment (or the link to the domain knowledge related to the general perception of the environment) etc. To be noted that the information layer of g_{perc} may contain more complex information related to the scope of g_{perc}; here we discuss only a simplified version of g_{perc}.

2. Let g_{plan} be another ic-granule, may be called the planner ic-granule, having in its information layer the specification of the general goal of the computation

at time t_0. The scope of g_{plan} corresponds to the particular module in the hard-drive of the robot where the goal description is set. The soft_suit, link_suit and the hard_suit of g_{plan} can be considered as different layers of that module. The reachability to more detailed and deeper layers of the plan in the hard_suit of the module happens through activating a decomposition of the plan from the directly reachable layer in the soft_suit.

3.   The ic-granule g_{kb} represents the knowledge base at time t_0. Here, g_{kb} can be considered as the module embedded in the hard drive of the robot where the domain knowledge is being represented. The information layer of g_{kb} consists of different relevant properties of different fragments of the robot and its surrounding. The soft_suit of g_{kb} contains different folders segregating the domain knowledge and the addresses to these folders are attached to the information layer of g_{kb}; in order to access the detailed information the files lying in the specific folders are to be opened following the addresses which are basically to locate the files. To be noted that the path for accessing the files through the addresses of the folder is metaphoric to the idea of accessing the hard_suit of g_{kb} through the information attached to its soft_suit.

So, the computation at time t_0 over the c-granule g_{perc} is led by the control at time t_0, which is given as follows.

$$Cont_{t_0}(g_{perc}) = (Nt_{t_0}, Lan_{upp}, Lan_{Low}, \mathcal{R}_{inf}, \mathcal{R}_{imp}, Mod_{imp})$$

The network of ic-granules, which are active at this stage, is given by $Nt_{t_0} = \{g_{perc}, g_{plan}, g_{kh}\}$. Based on the information gathered from the information layers of g_{perc}, g_{plan} and g_{kb}, the control with the help of $\mathcal{R}_{inf}$ decomposes a more detailed plan of actions, and then $\mathcal{R}_{imp}$ translates this plan of actions in the language of actuators. Here to be noted, that this plan of actions may be related to performing more measurements for gaining a better perception of the environment, or performing some other actions to reach to a desired state of the environment, or simply related to performing actions for achieving the goal. In general, a sequence of above mentioned performances may be required to better understand the currently perceived situation and in turn achieve the target goals. Each such actions must be realized by the module Mod_{imp}, and after realization of the selected action(s) the configuration of the ic-granules in Nt_0 may get changed.

Computation at time t_1: After the decomposition of the plan of action in the abstract language and then in the implementational level language, it starts the next cycle of the computation, may be marked as computation at time t_1. With the decomposition of the initial goal, available at the planner ic-granule g_{plan} at t_0, a new ic-granule is manifested with the label of the new detailed plan. Let us call this new planner ic-granule at time t_1 as g_{plan}^1. From the perspective of the example, at t_0 the information attached to g_{plan} encodes the general goal of the robot that primarily gets set in its hard-drive, i.e., in the soft_suit of g_{plan}. At time t_0 the hard_suit of g_{plan}, such as more underlying folders with more decomposed plans, remains still unaccessed. Based on the collected information of g_{perc} and g_{kb}, at time t_0 the control of the robot explores the rules from $\mathcal{R}_{inf}$. As the rules from $\mathcal{R}_{inf}$ help to determine the plan of actions in an abstract level language through this exploration the control of the robot initiates an interaction to the hard_suit of g_{plan} in order to get access to a more decomposed plan suitable for the current perception of the environment. Accordingly, $\mathcal{R}_{imp}$ translates the respective plan of actions in the language of the actuators of the robot.

1. From the perspective of our example, when the plan of actions is translated in the language of the actuators, like hands, legs, and the stick of the robot, a new ic-granule is manifested at this layer. Let us call it as g_{imp}^1, an implementational ic-granule. To be noted that g_{imp}^1 does not concern about the actual actuators; rather it is like another hard-drive in the implementational module Mod_{imp} of the hard-drive of the robot where the action plan can be stored in the language of actuators. The information layer of g_{imp}^1 also contains the specification of the conditions for initiating the implementation plan through the real actuators.

2. So, at time t_1 the network of ic-granules, which are actively participating in the computation led by the control, consists of the previous ic-granules g_{perc}, g_{kb}, and the newly generated ic-granules g_{plan}^1 and g_{imp}^1.

3. The plan of implementation as well as the condition of implementation are available at the information layer of g_{imp}^1; following that the real physical actions need to be physically performed by the robot through the functionalities of the actuators in the Mod_{imp} module. This real physical performance of actions is beyond mathematical formulation, and the model of IGrC only can perceive the result of performance of the actions in the real physical world after some time, say at time t_2. This marks the next cycle of the computation.

So, control at time t_1 can be presented as follows.

$$Cont_{t_1}(g_{perc}) = (Nt_{t_1}, Lan_{upp}, Lan_{low}, \mathcal{R}_{inf}, \mathcal{R}_{imp}, Mod_{imp})$$

The network of ic-granules at this stage is represented by $Nt_{t_1} = \{g_{perc}, g_{kb}, g_{plan}^1, g_{imp}^1\}$ where g_{imp}^1 is the implementational ic-granule conducting the real physical interactions through the implementational module Mod_{imp}.

Computation at time t_2 to t_n for some finite n: The specification of the plan of implementation of g_{imp}^1 is now realized through some physical objects during the time interval $[t_1, t_2]$. In case of the example, this object can be the stick of the robot on which the implementation plan is embedded, and thus g_{stick} represents the concerned c-granule, which is a sub-c-granule of the original ic-granule g_{perc}. At time t_2 after implementation of real physical actions the configuration of g_{perc} may change. Let us denote this changed ic-granule at time t_2 as g_{perc}^1.

1. After any real physical interaction with the physical reality the control starts with a new cycle starting by perceiving the environment and updating the perception based information. So, at time t_2 with the help of the information

layers of the ic-granules g^1_{plan}, g^1_{imp}, g^1_{perc} and g_{kb} the relation $\mathcal{R}_{inf}$ generates an abstract plan for perceiving the (not yet perceived part of) environment of g^1_{perc} and the relation $\mathcal{R}_{imp}$ generates an implementation plan for realizing the process of perception. The control at time t_2 can be represented as follows.

$Cont_{t_2}(g^1_{perc})$ can be regarded as $(Nt_{t_2}, Lan_{upp}, Lan_{low}, \mathcal{R}_{inf}, \mathcal{R}_{imp}, Mod_{imp})$

The network of ic-granules at this stage is $Nt_{t_2} = \{g^1_{perc}, g_{stick}, g_{kb}, g^1_{plan}, g^1_{imp}\}$.

2. Through the module Mod_{imp} implementation of the perception process runs over a time interval, say $[t_2, t_3]$, and the next cycle of computation starts at time t_3 with the new plan of actions in the ic-granules g^2_{plan} and g^2_{imp} respectively.

3. As g_{stick} is a sub-c-granule of the ic-granule g_{perc} some possible physical interactions of the stick with other objects in g_{perc} and the respective outcomes are encoded in the information layer of g_{perc} and thus can be used by the control. For example, the information may describe such rules that if the stick hits to a stone, there is a vibration. So, after perceiving the environment at time t_3 the information layer of g^1_{perc} as well as its configuration of objects may change, and a new ic-granule g^2_{perc} may be manifested. If the properties of perception recorded at g^2_{perc} matches to (a significant degree) the properties encoded as the expected outcomes of the interactions of g_{stick} with other objects in the scope of g^2_{perc}, the implementation of the plan of actions by the robot is considered to be successful; otherwise based on the information layers of g^2_{perc}, g_{kb}, g^2_{plan}, g^2_{imp}, the control $Cont_{t_3}(g^2_{perc})$ may generate the next plan of actions with the help of $\mathcal{R}_{inf}$ and $\mathcal{R}_{imp}$.

4. Here to be noted, that the generation of new action plans, based on the currently perceived situation and the goal of the computation, can continue for finitely many time points. Thus, at each stage the perception of the environment may get closer to the expected target environment through manifestation of several ic-granules g_{perc}, g^1_{perc}, g^2_{perc}, ... g^n_{perc}. Figure 3.2 presents the computation process running over a finite number of time points.

We can see that at each stage when a new ic-granule g^i_{perc} $(1 \leq i \leq n)$ is generated, a new planner ic-granule g^i_{plan} is also generated to specify what is to be done with the newly generated ic-granule. This sequence of planner ic-granules generated from time t_0 to the end, namely g_{plan}, g^1_{plan}, ..., g^i_{plan}, ... g^n_{plan} satisfies some conditions; the most basic of these conditions can be stated as follows.

(P1) Each g^i_{plan}[3] must be a sequence of g^{i1}_{plan}, ... g^{ik}_{plan} where $i < i1, ..., ik \leq n$.

(P2) For g_{plan} the last member of the sequence, say g^{0k}_{plan} must be the same, or at least significantly close[4] to g^n_{plan}.

---

[3] g_{plan} can be read as g^0_{plan}.

[4] In what sense a plan can be significantly close to another plan is not discussed here, but of course such technicalities can be introduced based on the context and purpose of the problem.

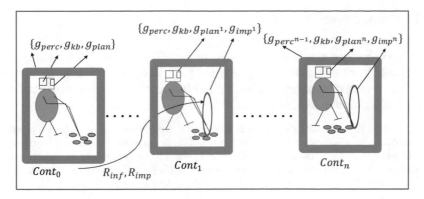

Fig. 3.2 Progress of the computation over the ic-granule containing the robot with a stick and its surrounding environment. At time t_0 the control has only the ic-granules g_{perc}, g_{kb} and g_{plan} as the active ic-granules. With the passage of time more detailed plan is generated and the implementational ic-granule g_{imp}^1 is generated to implement the abstract specification of the plan of actions through the language of the actuators. This process of generation of new ic-granules with more decomposed plan and implementation of that plan is repeated for finitely many time points t_0 to t_n

So, the computation process may stop when the g_{plan}^n is realized and its closeness to g_{plan}^{0k} is verified. Basically, within this sequence of planner ic-granules, in a regular interval the closeness of reaching a sub-goal g_{plan}^i is verified by the control using the information layer of g_{plan}^i and g_{perc}^{i+1}.

Here, in the example we have mentioned about the decomposition of the plan of actions from the initial stage g_{plan} to a stage g_{plan}^n, where at each stage it gets translated to a respective implementational level language. However, in practice decomposition of the action plan g_{plan} may have several layers of decompositions in between, as some plans may be launched based on the current perceptions, while the other ones may be related to the main goal. Moreover, apart from the two basic conditions (P1) and (P2), based on the context of the problem there can be different variant conditions as the stopping criterion of a computation running over a c-granule.

For example, for certain contexts, all the members of the sequence of the planner ic-granules $g_{plan}, g_{plan}^1, \ldots, g_{plan}^n$ must satisfy a degree of similarity, which can be set before. If at a stage of the computation a planner ic-granule g_{plan}^i crosses the pre-fixed degree of similarity the process may halt, indicating an infeasible path has been followed towards realizing the initial goal. This can be considered as an example of such contexts where much deviation from the initial plan may bring some risks. Instead of the above possibility one can consider also such situations where a pre-fixed threshold as the degree of similarity among $g_{perc}, g_{perc}^1, \ldots, g_{perc}^n$ is set. This may indicate those contexts where much deviation from the environment of the computation may bring risks. Plenty of such possibilities may arise, and we keep such considerations for future possible extension of the work.

3.4 Conclusion

The idea of computation over a c-granule, as described in Sect. 3.3, is in the line of the thought that Luc Steels has expressed in [20], according to which the complex dynamical systems (complex systems, for short) are considered as systems consisting of a set of interacting elements

> [...] *where the behavior of the total is an indirect, non-hierarchical consequence of the behavior of the different parts. Complex systems differ in that sense from strictly hierarchical systems [...] where the total behavior is a hierarchical composition of the behavior of the parts. In complex systems, global coherence is reached despite purely local non-linear interactions. There is no central control source. Typically the system is open.*

From the discussed example of the process of computation over a c-granule, it is quite clear that the process of computation deals with a network of hunks [12] of real physical matter associated with their information layers. The information layers indicate where, when and how they can be accessed, properties about them can be achieved and verified. So, one can notice how the notion of c-granule itself connect between the abstract world, that is the information layer, and the real physical semantics, that is the three-layered hunk of objects.

Referring to the above quotation, it is to be noticed that in the context of a computation over c-granule, as exemplified in Sect. 3.3.2, the control is a varying notion because over the passage of time the network of active granules changes from one to the other. The phrase *local non-linear interactions* also suits appropriately to the context of computation over a c-granule. From the example discussed in Sect. 3.3.2 we have seen how interactions made in g_{kb} help to prescribe a plan of actions and how interactions made in g_{imp} lead to revise or follow the previously made plan of actions. All these interactions are localized in the respective ic-granules but they together influence the path of the computation. The hierarchical layers of the computation may be marked by the changes in the configurations of the ic-granules from the networks of ic-granules in the respective controls at consecutive time points. Here to be noted as well, that these networks of ic-granules from one layer to the next layer may not be ordered in a linear manner as the changes in the configurations of the ic-granules from one layer to another may indicate complete and/or partial removal and/or addition of an ic-granule from one layer to the next layer. Moreover, as in IGrC the model allows a space for the real physical implementations by lifting the static description of a process to the level of real physical actions, the phrase *system is open*, from the quotation, also applies to it.

The implementational module Mod_{imp}, in the control of a c-granule, creates a specific interface between the abstract and the physical world. In particular, when the abstract specification of the plan of actions gets translated in the language of implementation, the control of the c-granule transits from the abstract module to the implementational module Mod_{imp}, which is supposed to initiate the real physical actions. These actions are not from the abstract mathematical space and the model of IGrC keep those action functions free from mathematical formalizations. The syntactic descriptions of those implementations as well as the expected properties of

Fig. 3.3 Interacting abstract
(AM) and physical (PM)
modules in control of
c-granules

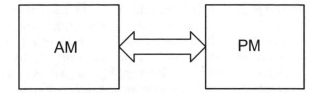

the real physical environment, after the implementation of the actions, are specified
in the informational layers of the implementational ic-granules belonging to the
control. The performance quality of those implemented actions can be verified with
the expected outcome by perceiving the changes in the world after initiation of the
actions. Thus the IGrC model keeps the possibility of mismatch between expected
and real physical outcome open as one never can a priori formalize all possible
outcomes of an action, which is supposed to be initiated in the physical world based
on the specification of the action plan, both in the abstract as well as lower level
language.

Moreover, the transition from one configuration of ic-granules to the other by
generating different ic-granules and using their information layers is not also purely
mathematical. Usually, in standard mathematical theories a state transition relation is
presented by a given family of sets $\{X_i\}_{i \in I}$ where a transition relation is represented as
a relation $tr_i \subseteq X_i \times X_i$. Here going beyond a mathematical formulation one needs
to incorporate (i) how elements of X_i are perceived in the real physical environment,
and (ii) how the transition relation tr_i is implemented in the real physical world. These
can be achieved by the control of a c-granule. The structure of a control is designed
to have two interacting modules, called the abstract module (AM) and the physical
module (PM) (see Fig. 3.3). As presented in Definition 3.1, AM pertains to the
information layers of the network Nt_i of ic-granules, two languages Lan_{upp}, Lan_{low},
and two relations $\mathcal{R}_{inf}$, $\mathcal{R}_{imp}$; whereas PM relates to the implementational module
Mod_{imp}. The communications between AM and PM creates the link between the
abstract and the physical world. Though the behaviour of AM and communication
of AM with PM can be represented in the framework of a mathematical modeling,
PM being composed of real physical objects lies outside of the abstract mathematical
modeling. However, partial information about properties of these objects and their
interactions may be communicated to AM through interaction of PM with AM.
Thus, IGrC endorses the view that complex phenomenon cannot be modelled in a
complete mathematical framework, and incorporates a model which going beyond
a pure mathematical modeling can keep constant connection with the ground of the
reality.

Addressing and convincing the need for such a new kind of modeling in the
context of artificial intelligence is a big challenge, and even bigger challenge is to
develop a general model which can encompass all possible real life phenomenon in
the above mentioned sense of computing. So, for now we attempt to only pen down
the features and factors of such a notion of computation, and present some of its
crucial components in a formal manner. As some of its components are grounded in

the real physical world, formalizing all the aspects of interactive granular computing in a pure mathematical framework perhaps would not be an easy task.

Why we require a three dimensional hunk of real physical objects labelled by an information layer may be explained better considering the analogy of writing a computer program in a language keeping a desired goal in mind, and having an outcome after running the program in a device. Writing a program can be thought of as specifying the plan of actions in an abstract language. While compiling or running it means translation of the abstract code in a machine level language. These are two different interfaces, each having its own formal set up; but a correctly written piece of code cannot display a correct output after compilation unless all the functionalities of the device, on which the compilation of the program is embedded, are working. That the device is functioning well, is outside of our formal mathematical world. Existing theories of AI only concern about separately designing these two interfaces assuming that the device is all set to take the input and show the output. Thus, there can be a huge gap between the expected outcome and the real outcome. AI from the perspective of IGrC considers, that apart from designing the two interfaces, namely the abstract and the implementational interfaces, the possibility of perceiving how the implementational interface is functioning in the reality, by a continuous process of verification, also need to be added in the system in order to simulate the behaviour of an autonomous agent.

Let us conclude the paper once again with a quotation which reflects a similar thought that simulating behaviour of a complex phenomenon in a complex system needs a new field of research which cannot be confined only in mathematical constructs. The area of research, which emphasizes that such a model must contain a harmonious symbiosis among things, data, and system, is known as Wisdom Web of Things (W2T) [21].

[Hyper world] *consists of the cyber, social, and physical worlds, and uses data as a bridge to connect humans, computers, and things.* [Wisdom Web of Things] *W2T focuses on the data cycle, namely 'from things to data, information, knowledge, wisdom, services, humans, and then back to things'. A W2T data cycle system is designed to implement such a cycle, which is, technologically speaking, a practical way to realize the harmonious symbiosis of humans, computers, and things in the emerging hyper world.*

Acknowledgements The author is very grateful to Prof. Andrzej Skowron for sharing and discussing his views related to the issues relevant to this work.

References

1. Ahn, H.S.: Formation control approaches for distributed agents. In: Studies in Systems, Decision and Control, vol. 205. Springer, Cham, Switzerland (2020). https://doi.org/10.1007/978-3-030-15187-4
2. Barsalou, L.W.: Perceptual symbol systems. Behav. Brain Sci. **22**(2), 577–660 (1999). https://doi.org/10.1017/S0140525X99002149

3. Brentano, F.: Psychologie vom empirischen Standpunkte. In: Studies in Systems, Decision and Control, vol. 205. Dunker & Humboldt, Leipzig (1874). https://archive.org/details/psychologievome02brengoog/page/n4/mode/2up
4. Brooks, F.P.: The mythical man-month: Essays on software engineering. Addison-Wesley, Boston (2020). (extended Anniversary Edition in 1995)
5. Deutsch, D., Ekert, A., Lupacchini, R.: Machines, logic and quantum physics. Neural Computation **6**, 265–283 (2000). https://doi.org/10.2307/421056
6. Dourish, P.: Where the action is. The foundations of embodied interaction. MIT Press, Cambridge, MA (2004)
7. Dutta, S., Skowron, A.: Interactive granular computing connecting abstract and physical worlds: An example. In: Schlingloff, H., Vogel, T. (eds.) In Proceedings of the 29th International Workshop on Concurrency, Specification and Programming CS & P 2021, CEUR Workshop Proceedings, pp. 46–59 (2021). http://ceur-ws.org/Vol-2951/
8. Dutta, S., Skowron, A.: Interactive granular computing model for intelligent systems. In: Shi, Z., Chakraborty, M., Kar, S. (eds.) Intelligence Science III. 4th IFIP TC 12 International Conference, Durgapur, India, February 24–27, 2021, Revised Selected Papers. IFIP Advances in Information and Communication Technology (IFIPAICT) book. ICIS2020, 623, pp. 37–48. Springer, Cham, Switzerland (2021). https://doi.org/10.1007/978-3-030-74826-5_4
9. Dutta, S., Skowron, A.: Toward a computing model dealing with complex phenomena: Interactive granular computing. In: Nguyen, N.T., Iliadis, L., Maglogiannis, I., Trawiński, B. (eds.) Computational Collective Intelligence. ICCCI 2021, LNCS, vol. 12876, pp. 199–214. Springer, Cham (2021). https://doi.org/10.1007/978-3-030-88081-1_15
10. Gerrish, S.: How smart machines think. MIT Press, Cambridge, MA (2018)
11. Harnad, S.: The symbol grounding problem. Physica **D42**, 335–346 (1990). https://doi.org/10.1016/0167-2789(90)90087-6
12. Heller, M.: The ontology of physical objects. Four dimensional hunks of matter. Cambridge Studies in Philosophy. Cambridge University Press, Cambridge, UK (1990)
13. Hodges, A.: Alan turing, logical and physical. In: Cooper, S. B. L (eds.) New Computational Paradigms. Changing Conceptions of What is Computable
14. Jankowski, A.: Interactive granular computations in networks and systems engineering: A practical perspective. Lecture Notes in Networks and Systems. Springer, Heidelberg (2017). https://doi.org/10.1007/978-3-319-57627-5
15. Nöe, A.: Action in perception. MIT Press, Cambridge (2004)
16. Ortiz, C., Jr.: Why we need a physically embodied turing test and what it might look like. AI Mag. **37**, 55–62 (2016). https://doi.org/10.1609/aimag.v37i1.2645
17. Pylyshyn, Z.W.: Computation and cognition. Toward a foundation for cognitive science. The MIT Press, Cambridge, MA (1884)
18. Skowron, A., Jankowski, A.: Rough sets and interactive granular computing. Fundam. Inf. **147**, 371–385 (2016). https://doi.org/10.3233/FI-2016-1413
19. Skowron, A., Jankowski, A., Dutta, S.: Interactive granular computing. Granular Comput. **1**(2), 95–113 (2016). https://doi.org/10.1007/s41066-015-0002-1
20. Steels, L.: The synthetic modeling of language origins. Evol. Commun. **1**(1), 1–34 (1997). https://doi.org/10.1075/eoc.1.1.02ste
21. Zhong, N., Ma, J.H., Huang, R.H., Liu, J.M., Yao, Y.Y., Zhang, Y.X., Chen, J.H.: Research challenges and perspectives on wisdom web of things (w2t). J. Supercomput. **64**, 862–882 (2013). https://doi.org/10.1007/s11227-010-0518-8

Chapter 4
Extended Future in Testing Semantics for Time Petri Nets

Elena Bozhenkova and **Irina Virbitskaite**

Abstract Dense-Time Petri Nets (TPNs) are a widely-accepted model suitable for qualitative and quantitative modeling and verifying safety-critical, computer-controlled, and real-time systems. Testing equivalences used to compare the behavior (processes) of systems and reduce their structure are defined in terms of tests that processes may or must pass. The intention of the paper is to present a framework for developing, studying and comparing testing equivalences with extended tests (extended future) in interleaving, partial order, and causal tree semantics in the context of safe TPNs (transitions are labeled with time firing intervals, can fire only if their lower time bounds are attained and must fire when their upper time bounds are reached). Additionally, we establish, for the whole class of TPNs and their various subclasses, relationships between testing equivalences and other equivalences from the interleaving—partial order and linear time—branching time spectra. This allows studying in complete detail the timing behavior, in addition to the degrees of relative concurrency and nondeterminism of processes.

4.1 Introduction

An essential constituent of any theory of systems is a notion of equivalence between the systems. In general, equivalences are useful in the setting of specification and verification with a view to both comparing distinct systems and reducing the structure of a system without affecting its behavior. By abstracting from undesirable details, behavioral equivalences serve to identify the systems that, when functioning, offer similar opportunities for interaction with the environment. Since the properties potentially relevant for the verification of concurrent systems are numerous, the behavioral equivalences promoted in the literature are many; the relationships

E. Bozhenkova (✉) · I. Virbitskaite
A.P. Ershov Institute of Informatics Systems, SB RAS, 6, Acad. Lavrentiev avenue, 630090 Novosibirsk, Russia
e-mail: bozhenko@iis.nsk.su

I. Virbitskaite
e-mail: virb@iis.nsk.su

© The Author(s), under exclusive license to Springer Nature Switzerland AG 2023
B. Schlingloff et al. (eds.), *Concurrency, Specification and Programming, Studies in Computational Intelligence 1091*, https://doi.org/10.1007/978-3-031-26651-5_4

between them have been fairly well-understood (see [10, 14] among others). In Petri net theory, the hierarchies of a wide range of behavioral equivalences in diverse semantics have been constructed for the whole class of the model and its subclasses (see [17, 19, 20] and references therein). As a rule, with a view to distinguishing the approaches, the comparision of behavioral equivalences for concurrent systems is based on the combination of two dichotomies.

The first dichotomy is the *interleaving—partial order spectrum*. The essence of the *interleaving* approach is to represent the concurrency relation as a nondeterministic choice between the executions of linearly ordered actions, rather than directly. Interleaving semantics of the Petri net is usually represented in terms of the sequences of transition firings in order to move from one net marking to another. The advantages of the interleaving interpretation of concurrency are its simplicity, mathematical elegance, and allowing for the analysis of some safety and liveness properties of systems. However, the difference between concurrency and nondeterminism is lost. A way to overcome the limitations of the interleaving approach is to model concurrency as a lack of causality dependence between system actions, which is presented, as a rule, by a *partial order*. Partial order semantics of the Petri net is most often represented by means of the so-called causal net processes consisting of causal nets and their homomorphisms to the Petri net (see [11, 16] and references therein). The causal net includes events (corresponding to transition firings) and conditions (corresponding to markings) connected by causality and concurrency. Partial order semantics is more suitable for studying such properties of concurrent systems as the absence of deadlocks, 'fairness', maximum parallelism, etc. In between the opposite sides of the spectrum under consideration, there is *causal tree* semantics based on a combination of interleaving and partial order. The behavior of the system is represented in the form of a tree with nodes represented by the interleaving sequences of executed actions and edges labelled by the corresponding actions and information about their partial order predecessors. Therefore, information about causality is kept precisely. Relationships between partial order and causal tree semantics for event structure models were thoroughly investigated in [9, 12]. To the best of our knowledge, causal tree semantics has not been studied yet in the framework of Petri nets.

The second dichotomy is the so-called *linear-time—branching-time spectrum*. Here, one looks at the various possibilities of taking into account the choice points between the alternative executions of systems. A conventional representative of the linear-time approach is *trace equivalence*, comparing the behaviors of systems in terms of their languages—the sets of possible executions of the systems; thus, the information about the choice points between alternative executions is completely ignored. In interleaving semantics, an execution of the Petri net is considered as a sequence of transition firings labeled by the corresponding actions, while in partial order semantics it is considered as a labeled causal net process generated during the operation of the Petri net. Two Petri nets are assumed interleaving (partial order) trace equivalent if their interleaving (partial order) languages coincide. The standard example of branching time equivalence is *bisimulation*, assuming that systems should behave in the same way, mutually simulating each other. Bisimulation carefully takes into account the branching structure of the behavior of systems. Furthermore, the formulation of bisimilarity is mathematically stylish and has received much attention in many fields of computer science. Two Petri nets are interleaving (partial order) bisimilar if each of the same interleaving (partial order) executions is recursively followed by the same actions. Here, two variants of the continuation of the Petri net executions at the next step are distinguished—one-action and many-actions extensions. The culminating point of the partial order approach to equivalences is history preserving bisimulation, originally proposed in terms of behavior structures in [18]. For transition-labeled Petri nets, E. Best and others [4] have shown the coincidence of both variants of the extensions with both interleaving bisimulation and history preserving bisimulation. It is worth to notice that some researchers consider bisimulation equivalence too discriminatory: two processes may be related, even if there is no way to establish this fact by bisimularity. Therefore, in between the opposite ends of the linear-time—branching-time spectrum, there is *testing equivalence*. It is explicitly based on a framework of extracting information about the system behaviors by testing them with the sets of subsequent actions. Two Petri nets are interleaving (partial order) testing equivalent if after each of the same interleaving (partial order) executions, called experiments, the nets can execute at least one action from the same sets of subsequent actions, called tests. To the best of our knowledge, the many-actions variant of tests has not been investigated in the setting of testing equivalence.

The safety of modern information and computer systems is becoming crucially important as their failure can cause huge losses, destruction, and even human casualties. To verify the behavior of these systems, we need formalisms making it possible to describe and analyze both the functional (qualitative) and real-time (quantitative) properties of the systems. Petri net theory offers various real-time extensions associating time characteristics with various net elements (places, transitions, arcs, and tokens) [6]. The most popular extension is dense-time Petri nets (TPNs) [13], where each transition has its local clock and static interval of real numbers (firing delays of the transition); the transition is herewith assumed to fire instantaneously. TPNs are known to be equipotent to Turing machines and include various classes of time extensions of Petri nets, which explains why the TPN model is of greater interest to researchers.

In [3], interleaving trace and bisimulation equivalences were used to compare the expressiveness of timed automata and time Petri nets. In [5], interleaving testing relations along with their alternative characterization and discretization extend to time Petri nets, where time characteristics are associated with tokens and time intervals are attached to arcs from places to transitions. Partial order semantics for safe TPNs was put forward in [2]. However, to the best of our knowledge, there have been only few mentions of a fusion of timing and partial order semantics in the behavioral equivalence scenario. In this regard, the paper [1] is a welcome exception. It contributes to the classification of a wealth of equivalences of linear time—branching time spectrum presented in interleaving, causal tree and partial order semantics of dense-time event structures, with and without internal actions. In [8, 21], the authors studied the relationships between trace and bisimulation equivalences in interleaving and partial order semantics of safe TPNs. Also, our origin is the paper [7], the main result of which is the coincidence of partial order and causal tree testing equivalences with tests containing the one-action extensions of experiments, in the setting of safe TPNs.

The intention of the paper is twofold. First, we define, study, and compare the one-action and many-actions variants of testing equivalences in interleaving, partial order and causal tree semantics in the context of safe TPNs whose transitions are labeled with time firing intervals; enabled transitions are able to fire only if their lower time bounds are attained and are forced to fire when their upper time bounds are reached. Second, we establish, for the whole class of TPNs and their subclasses, the place of these variants of testing equivalences among other equivalences from the interleaving—partial order and linear time—branching time spectra.

The paper is organized as follows. The next section provides the syntax and various (interleaving, causal tree, and partial order) semantics of TPNs. In Sect. 4.3, we introduce and study the one-action and many-actions variants of trace, testing and bisimulation equivalences in the semantics under consideration. In the following section, hierarchies of the equivalences interrelations are developed for the entire class of TPNs and their subclasses. Section 4.5 concludes the paper.

4.2 Syntax and Different Semantics of TPNs

In this section, some terminology concerning the syntax of Petri nets with timing constraints (time intervals on the transition firings [2]) and their semantics in terms of firing sequences (schedules), causal net processes and causal trees are defined.

4.2.1 Syntax and Interleaving Semantics of TPNs

We start with recalling the definitions of the structure and interleaving behavior of time Petri nets (TPNs). The TPN consists of an underlying Petri net and a static timing function mapping each transition to a time interval with non-negative rational boundaries. Each enabled transition is assumed to have its own local clock. A marking alone is not enough to describe the state of the TPN, so a dynamic timing function is added to indicate the clock values—the time elapsed since the more recent dates at which the associated transitions became enabled. The initial state consists of the initial marking and dynamic timing function with zero clock values for all enabled transitions. A transition can fire from a current state only if the transition is enabled at the current marking and its clock value belongs to its time interval. The firing of a transition which can fire from a state after a delay time results in a new state, i.e. a new marking and new dynamic timing function. Its clock values are reset to zero for the newly enabled transitions, and its old clock values are increased by the delay time for the transitions which continue to be enabled. A sequence of changes in states is called a firing schedule of the TPN. The firing schedules from the initial state represent interleaving semantics of the TPN.

We use *Act* as an alphabet of actions. Let the domain $\mathbb{T}$ of time values be the set of non-negative rational numbers. We denote by $[\tau_1, \tau_2]$ the closed interval between two time values $\tau_1, \tau_2 \in \mathbb{T}$. Infinity is allowed at the upper bounds of the intervals. Let *Interv* be the set of all such intervals.

Definition 4.1 • A *(labeled over Act) time Petri net* is a pair $\mathcal{TN} = (\mathcal{N}, D)$, where $\mathcal{N} = (P, T, F, M_0, Abs)$ is a (labeled over *Act*) underlying Petri net (with a finite set P of places, a finite set T of transitions such that $P \cap T = \emptyset$ and $P \cup T \neq \emptyset$, a flow relation $F \subseteq (P \times T) \cup (T \times P)$, an initial marking $\emptyset \neq M_0 \subseteq P$, a labeling function $Abs : T \to Act$) and $D : T \to Interv$ is a *static timing function* associating with each transition a time interval. For a transition $t \in T$, the boundaries of the interval $D(t) \in Interv$ are called the *earliest firing time* Eft and *latest firing time* Lft of t. For $x \in P \cup T$, let $^\bullet x = \{y \mid (y, x) \in F\}$ and $x^\bullet = \{y \mid (x, y) \in F\}$ be the *preset* and *postset* of x, respectively. For $X \subseteq P \cup T$, define $^\bullet X = \bigcup_{x \in X} {}^\bullet x$ and $X^\bullet = \bigcup_{x \in X} x^\bullet$.

• A marking M of $\mathcal{TN}$ is any subset of P. A transition $t \in T$ is *enabled at a marking* M if $^\bullet t \subseteq M$. Let $En(M)$ be the set of transitions enabled at M. A *state* of $\mathcal{TN}$ is a pair (M, I), where M is a marking and $I : En(M) \longrightarrow \mathbb{T}$ is a *dynamic timing function*. The *initial state* of $\mathcal{TN}$ is a pair $S_0 = (M_0, I_0)$, where M_0 is the initial marking and $I_0(t) = 0$, for all $t \in En(M_0)$. A transition t enabled at a marking M *can fire from a state* $S = (M, I)$ *after a (relative) delay time* $\theta \in \mathbb{T}$ if $(Eft(t) \leq I(t) + \theta)$, and $(I(t') + \theta \leq Lft(t'))$, for all $t' \in En(M)$. The *firing* of a transition t that can fire from a state $S = (M, I)$ after a delay time θ *leads to* the new state $S' = (M', I')$ (denoted $S \xrightarrow{(t,\theta)} S'$) given by:

(a) $M' = M \setminus {}^\bullet t \cup t^\bullet$,

(b) $\forall t' \in T \circ I'(t') = \begin{cases} I(t') + \theta, & \text{if } t' \in En(M \setminus {}^\bullet t), \\ 0, & \text{if } t' \in En(M') \setminus En(M \setminus {}^\bullet t), \\ \text{undefined}, & \text{otherwise.} \end{cases}$

We write $S \xrightarrow{(a,\theta)} S'$, if $S \xrightarrow{(t,\theta)} S'$ and $a = Abs(t)$. We use the notation $S \xrightarrow{\sigma} S'$ iff $\sigma = (t_1, \theta_1) \ldots (t_k, \theta_k)$ and $S = S^0 \xrightarrow{(t_1,\theta_1)} S^1 \ldots S^{k-1} \xrightarrow{(t_k,\theta_k)} S^k = S'$ $(k \geq 0)$. In this case, σ is a *firing schedule of $\mathcal{TN}$ from S (to S')*, and S' is a *reachable state of $\mathcal{TN}$ from S*. Let $\mathcal{FS}(\mathcal{TN})$ be the set of all firing schedules of $\mathcal{TN}$ from S_0, and $RS(\mathcal{TN})$ be the set of all reachable states of $\mathcal{TN}$ from S_0. For firing schedules $\sigma = (t_1, \theta_1) \ldots (t_k, \theta_k)$ and $\sigma' = (t_1, \theta_1) \ldots (t_k, \theta_k) (t_{k+1}, \theta_{k+1}) \ldots (t_{k+l}, \theta_{k+l})$ from $\mathcal{FS}(\mathcal{TN})$, we call σ' is a *one-extension (many-extension)* of σ, if $l = 1$ $(l \geq 1)$. In addition, we write $Abs(\sigma) = (a_1, \theta_1) \ldots (a_k, \theta_k)$ iff $a_i = Abs(t_i)$, for all $1 \leq i \leq k$.

We call $\mathcal{TN}$ *T-restricted* iff $^\bullet t \neq \emptyset \neq t^\bullet$, for all transitions $t \in T$; *contact-free* iff whenever a transition t can fire from the state $S = (M, I)$ after some delay time θ, then $(M \setminus {}^\bullet t) \cap t^\bullet = \emptyset$, for all $S \in RS(\mathcal{TN})$. In what follows, we shall consider only T-restricted and contact-free time Petri nets.

Fig. 4.1 A time Petri net $\widetilde{T\mathcal{N}}$

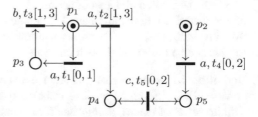

Example 1 The (labeled over $Act = \{a, b, c\}$) time Petri net $\widetilde{T\mathcal{N}}$ is shown in Fig. 4.1. Here, the net places are represented by circles and the net transitions—by bars; the names are depicted near the elements; the flow relation is drawn by arcs; the initial marking is represented as the set of the places with tokens (bold points); and the values of the labeling and timing functions are printed next to the transitions. It is not difficult to check that t_1, t_2 and t_4 are transitions enabled at the initial marking $M_0 = \{p_1, p_2\}$. These transitions can fire from the initial state $S_0 = (M_0, I_0)$ after the time delay $\theta_1 = 1$, where $I_0(t) = \begin{cases} 0, & \text{if } t \in \{t_1, t_2, t_4\}, \\ undefined, & \text{otherwise.} \end{cases}$ The sequence $\sigma = (t_1, 1)\ (t_4, 0)\ (t_3, 1)\ (t_2, 2)\ (t_5, 2)$ is a firing schedule of $\widetilde{T\mathcal{N}}$ from S_0. Also, we have $Abs(\sigma) = (a, 1)\ (a, 0)\ (b, 1)\ (a, 2)\ (c, 2)$. Furthermore, it is easy to see that $\widetilde{T\mathcal{N}}$ is really T-restricted and contact-free.

4.2.2 Causal Net Process Semantics of TPNs

In this subsection, the concept of causality-based net processes is presented in the context of time Petri nets.

We begin with considering the notion of a time causal net, the events and conditions of which are related by causality and concurrency (absence of causality), and whose timing function associates their (global) occurrence times with the events.

Definition 4.2 • A *(labeled over Act) time causal net* is a finite, acyclic net $TN = (B, E, G, Abs, \tau)$ with a set B of conditions; a set E of events; a flow relation $G \subseteq (B \times E) \cup (E \times B)$ such that $\mid {}^\bullet b \mid \leq 1 \geq \mid b^\bullet \mid$, for all $b \in B$, and ${}^\bullet B = E = B^\bullet$; a labeling function $Abs : E \to Act$, and a timing function $\tau : E \to \mathbb{T}$ such that $e\ G^+\ e' \Rightarrow \tau(e) \leq \tau(e')$. Let ${}^\bullet TN = \{b \in B \mid {}^\bullet b = \emptyset\}$.
- $\prec = G^+$, $\preceq = G^*$ (causality).
- For an event $e \in E$, $Predec(e) = \{e' \in E \mid e' \prec e\}$ (causal predecessors of e), $Earlier(e) = \{e' \in E \mid \tau(e') < \tau(e)\}$ (time predecessors of e), and $Cut(e) = (Earlier(e)^\bullet \cup {}^\bullet TN) \setminus {}^\bullet Earlier(e)$.
- E' is a *downward-closed subset* of E iff $E' \subset E$ and $Predec(e') \subseteq E'$, for all $e' \in E'$; a *timely sound subset* of E iff $E' \subset E$ and $Earlier(e') \subseteq E'$, for all $e' \in E'$.
- For events $e, e' \in E$, $e \smile e' \iff \neg((e \prec e') \vee (e' \prec e))$ (concurrency).

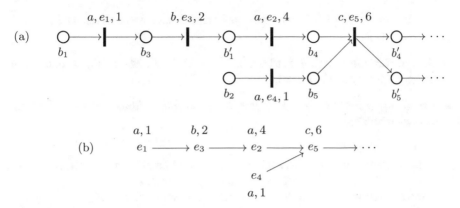

Fig. 4.2 **a** A time causal net $\widetilde{TN}$ and **b** its time poset $\eta(\widetilde{TN})$

- A sequence $\rho = e_1 \ldots e_k$ $(k \geq 0)$ of events of TN is a *linearization* of TN iff (i) every event of TN appears in the sequence exactly once and (ii) both causal and time order are preserved: for all $1 \leq i, j \leq k$, $(e_i \prec e_j) \vee (\tau(e_i) < \tau(e_j)) \Rightarrow i < j$.[1]
- $\eta(TN) = (E_{TN}, \preceq_{TN} \cap(E_{TN} \times E_{TN}), Abs_{TN}, \tau_{TN})$ is a *time poset*.[2] Let $\mathcal{TP}_{Act}$ be the set of all time posets labelled over Act.
- Time posets $\eta = (E, \preceq, Abs, \tau), \eta' = (E', \preceq', Abs', \tau') \in \mathcal{TP}_{Act}$ are *isomorphic* (denoted $\eta \simeq \eta'$) iff there is a bijective mapping $\beta : E \rightarrow E'$ such that (i) $e \preceq e' \iff \beta(e) \preceq' \beta(e')$, for all $e, e' \in E$; (ii) $Abs(e) = Abs'(\beta(e))$, for all $e \in E$, and (iii) $\tau(e) = \tau'(\beta(e))$, for all $e \in E$.
- For time posets $\eta = (E, \preceq, Abs, \tau)$, $\eta' = (E', \preceq', Abs', \tau') \in \mathcal{TP}_{Act}$, and $\star \in \{one, many\}$, η' is a $\star$-*extension of* η iff E is a downward-closed and timely sound subset of E', $\preceq=\preceq' \cap(E \times E)$, $Abs = Abs'\mid_E$, $\tau = \tau'\mid_E$, and $\mid E' \setminus E \mid = 1$, if $\star = one$.

Example 2 The time causal net $\widetilde{TN}$ is depicted in Fig. 4.2a. Here, the net conditions are represented as circles and net events—as bars; the flow relation is drawn by arcs; the names and, in addition, the values of the labeling and timing functions are printed next to the elements. Clearly, $\bullet \widetilde{TN} = \{b_1, b_2\}$. Besides, we have $Predec(e_2) = \{e_1, e_3\}$, $Earlier(e_2) = \{e_1, e_3, e_4\}$, $Cut(e) = (\{b_3, b_1', b_5\} \cup \{b_1, b_2\}) \setminus \{b_1, b_2, b_3\} = \{b_1', b_5\}$. It is easy to see that $\{e_1, \ldots, e_4\}$ is downward-closed and timely sound subset of $\{e_1, \ldots, e_5\}$. The event e_4 is concurrent with the events e_1, e_2, and e_3. The sequence $e_1e_4e_3e_2e_5$ is a linearization of $\widetilde{TN}$. The time poset $\eta(\widetilde{TN})$ is shown in Fig. 4.2b.

---

[1] The alternative definition of (ii): for all $1 \leq i < j \leq k$ it holds: $\bigcup_{1 \leq l \leq i} e_l$ is a downward-closed and timely sound subset of $\bigcup_{1 \leq m \leq j} e_l$.

[2] A *(labeled over Act) time poset (partially ordered set)* is a tuple $\eta = (X, \preceq, \lambda, \tau)$ consisting of a finite set X of elements; a reflexive, antisymmetric and transitive relation $\preceq$; a labeling function $l : X \rightarrow Act$; and a timing function $\tau : X \rightarrow \mathbb{T}$ such that $x \preceq x' \Rightarrow \tau(x) \leq \tau(x')$.

We are now ready to define the concept of causal net based processes of the TPN, proposed in [2].

Definition 4.3 Given a time Petri net $\mathcal{TN} = ((P, T, F, M_0, Abs), D)$ and a time causal net $TN = (B, E, G, l, \tau)$,

- a mapping $\varphi : B \cup E \to P \cup T$ is a *homomorphism* from TN to $\mathcal{TN}$ iff the following holds:

  - $\varphi(B) \subseteq P, \varphi(E) \subseteq T$;
  - the restriction of φ to $^{\bullet}e$ is a bijection between $^{\bullet}e$ and $^{\bullet}\varphi(e)$, and the restriction of φ to $e^{\bullet}$ is a bijection between $e^{\bullet}$ and $\varphi(e)^{\bullet}$, for all $e \in E$;
  - the restriction of φ to $^{\bullet}TN$ is a bijection between $^{\bullet}TN$ and M_0;
  - $Abs(e) = Abs(\varphi(e))$, for all $e \in E$.

- A pair $\pi = (TN, \varphi)$ is a *causal process of a time Petri net* $\mathcal{TN}$ iff TN is a time causal net and φ is a homomorphism from TN to $\mathcal{TN}$.
- A causal process $\pi = (TN, \varphi)$ of $\mathcal{TN}$ is *correct* iff for all $e \in E$ it holds:

(a) $\tau(e) \geq \mathbf{TOE}_\pi(^{\bullet}e, \varphi(e)) + Eft(\varphi(e))$,
(b) $\forall t \in En(\varphi(Cut(e))) \circ \tau(e) \leq \mathbf{TOE}_\pi(Cut(e), t) + Lft(t)$.

   Here, for a subset $B' \subseteq B_{TN}$ and a transition $t \in En(\varphi(B'))$, the time of enabling (TOE) of t, i.e. the latest global time moment when tokens appear in all input places of t, is defined as follows: $\mathbf{TOE}_\pi(B', t) = \max\left(\{\tau_{TN}(^{\bullet}b) \mid b \in B'_{[t]} \setminus {}^{\bullet}TN\} \cup \{0\}\right)$, where $B'_{[t]} = \{b \in B' \mid \varphi_{TN}(b) \in {}^{\bullet}t\}$.

   Let $\mathcal{CP}(\mathcal{TN})$ denote the set of correct causal processes of $\mathcal{TN}$. We assume that $\pi_0 = (TN_0 = (B_0, \emptyset, \emptyset, \emptyset, \emptyset), \varphi_0) \in \mathcal{CP}(\mathcal{TN})$ with $\varphi_0(B_0) = M_0$ is the initial causal process of $\mathcal{TN}$.
- Causal processes $\pi = (TN, \varphi)$, $\pi' = (TN', \varphi') \in \mathcal{CP}(\mathcal{TN})$ are *isomorphic* (denoted $\pi \simeq \pi'$) iff there exists an isomorphism $f : \eta(TN) \to \eta(TN')$ such that $\varphi(e) = \varphi'(f(e))$, for all $e \in E$.
- For causal processes $\pi = (TN, \varphi)$, $\pi' = (TN', \varphi') \in \mathcal{CP}(\mathcal{TN})$, and $\star = \{one, many\}$, π' is a $\star$-extension of π in $\mathcal{TN}$ iff $\eta(TN')$ is a $\star$-extension of $\eta(TN)$, and, moreover, $B \subseteq B', G = G' \cap (B \times E \cup E \times B)$, and $\varphi = \varphi' \mid_{B \cup E}$.

Example 3 A causal process of the TPN $\widetilde{\mathcal{TN}}$ depicted in Fig. 4.1 is a pair $\widetilde{\pi} = (\widetilde{TN}, \varphi)$, where $\widetilde{TN}$ is shown in Fig. 4.2a and $\varphi(e_i) = t_i$, for all $1 \leq i \leq 5$, $\varphi(b_i) = p_i$, for all $1 \leq i \leq 5$, $\varphi(b'_j) = p_j$, for all $j \in \{1, 4, 5\}$. Check that $(\widetilde{TN}, \varphi)$ is correct, for example, w.r.t. the event e_5. First, calculate $Cut(e_5) = \{b_4, b_5\}$, $En(\varphi(Cut(e_5))) = \{t_5\}$, $\mathbf{TOE}_{\widetilde{\pi}}(^{\bullet}e_5 = \{b_4, b_5\}, \varphi(e_5) = t_5) = \max\{\tau_{TN}(^{\bullet}b_4) = 4)$, $\tau_{TN}(^{\bullet}b_5) = 1\} = 4 = \mathbf{TOE}_{\widetilde{\pi}}(Cut(e_5) = \{b_4, b_5\}, t_5)$. We get that $\tau(e_5) = 6 \geq \mathbf{TOE}_{\widetilde{\pi}}(^{\bullet}e_5, \varphi(e_5)) = 4 + (Eft(\varphi(e_5)) = 0) = 4$, and $\tau(e_5) = 6 \leq \mathbf{TOE}_{\widetilde{\pi}}(Cut(e_5), t_5) = 4 + (Lft(t_5) = 2) = 6$. In a similar way, we can make sure that $(\widetilde{TN}, \varphi)$ is correct, i.e. correct w.r.t. the events $e_1, \ldots, e_5$.

For the time Petri net, we present the relationships between its correct causal processes and its interleaving firing schedules from the initial state.

Proposition 4.1 [7] Given a time Petri net $\mathcal{TN}$,

(i) *for any causal process $\pi = (TN, \varphi) \in \mathcal{CP}(\mathcal{TN})$ with any linearization $\rho = e_1 \ldots e_k$ of TN, there is a unique firing schedule $\sigma_{\pi,\rho} = FS_\pi(\rho) = (\varphi(e_1), \tau(e_1) - 0) \ldots (\varphi(e_k), \tau(e_k) - \tau(e_{k-1})) \in \mathcal{FS}(\mathcal{TN})$;*

(ii) *for any firing schedule $\sigma \in \mathcal{FS}(\mathcal{TN})$, there is a unique (up to an isomorphism) causal process $\pi_\sigma = (TN, \varphi) \in \mathcal{CP}(\mathcal{TN})$ with a unique linearization ρ_σ of TN such that $FS_{\pi_\sigma}(\rho_\sigma) = \sigma$.*

The facts established above lead us to

Corollary 4.1 *Given a time Petri net $\mathcal{TN}$,*

(i) *for any causal process $\pi = (TN, \varphi) \in \mathcal{CP}(\mathcal{TN})$ with any linearization ρ of TN, it holds that $\pi \simeq \pi_{\sigma_{\pi,\rho} = FS_\pi(\rho)}$;*

(ii) *for any firing schedule $\sigma \in \mathcal{FS}(\mathcal{TN})$, $\sigma = \sigma_{\pi_\sigma, \rho_\sigma}$.*

We expand Proposition 4.1 to the extensions of the correct causal processes of the time Petri net.

Lemma 4.1 *Given a time Petri net $\mathcal{TN}$ and $\star \in \{one, many\}$,*

(i) *for any causal process $\pi = (TN, \varphi) \in \mathcal{CP}(\mathcal{TN})$ with any linearization ρ of TN, whenever $\pi' = (TN', \varphi')$ is a $\star$-extension of π in $\mathcal{TN}$, then $\sigma_{\pi',\rho'}$ is a $\star$-extension of $\sigma_{\pi,\rho}$ in $\mathcal{TN}$, for all linearizations $\rho' = \rho\widehat{\rho}$ of TN';*

(ii) *for any firing schedule $\sigma \in \mathcal{FS}(\mathcal{TN})$, whenever σ' is a $\star$-extension of σ in $\mathcal{TN}$, then $\pi_{\sigma'}$ is a $\star$-extension of π_σ in $\mathcal{TN}$ and $\rho_{\sigma'} = \rho_\sigma \widehat{\rho}$.*

Proof Sketch (i) Using the definitions of the linearization of the time causal net and a $\star$-extension of the causal process of the TPN, we can obtain linearizations $\rho' = \rho\widetilde{\rho}$ of TN', for the given linearization ρ of TN. Then, thanks to the definition of the function FS, $\sigma_{\pi',\rho'}$ is a $\star$-extension of $\sigma_{\pi,\rho}$ in $\mathcal{TN}$, for all ρ' of TN'.

(ii) Immediately follows from the algorithm (see Theorem 21 in [2]) to construct for a firing schedule the causal process and from its uniqueness (up to isomorphism), due to Proposition 4.1(ii).   □

4.2.3 Causal Tree Semantics of TPNs

Causal trees [9] are synchronization trees carring additional information about the causes of actions in their edge labels, thereby providing us with an interleaving description of concurrent processes faithfully expressing causality. In the causal tree of the TPN, the vertices are the interleaving firing schedules of the TPN; there is an edge between two vertices if the second vertice is a one-action extension of the first one. The causes in the edge labels are determined based on the causality relations in the correct causal processes corresponding to the vertices (the firing schedules). Timing in the edge labels is represented by the delay times in the firing schedules.

Definition 4.4 • The *causal tree* of the TPN $\mathcal{TN}$ is a tree $CT(\mathcal{TN}) = (V = \mathcal{FS}(\mathcal{TN}), E, Abs)$, where V is the set of vertices with the root ϵ; $E = \{(\sigma, \sigma(t, \theta)) \mid \sigma, \sigma(t, \theta) \in V\}$ is the set of edges; Abs is the labeling function such that $Abs(\epsilon) = \epsilon$ and $Abs(\sigma, \sigma(t, \theta)) = (Abs_{\mathcal{TN}}(t), \theta, K)$, with $\sigma = FS_{\pi_\sigma}(\rho_\sigma = e_1 \ldots e_n)$, $\sigma(t, \theta) = FS_{\pi_{\sigma(t, \theta)}}(\rho_{\sigma(t, \theta)} = e_1 \ldots e_n e)$, $K = \{n - l + 1 \mid e_l \prec_{TN_{\pi_{\sigma(t, \theta)}}} e\}$.
- For a vertex $\sigma \in V$, $path(\sigma)$ is the path starting from the root and finishing in $\sigma \in V$.[3] Let $\mathcal{P}(CT(\mathcal{TN})) = \{path(\sigma) \mid \sigma \in V\}$ denote the set of all paths from the root in $CT(\mathcal{TN})$. We assume that $u_0 = path(\epsilon) = \epsilon$ is the initial path in $CT(\mathcal{TN})$.
- For paths $path(\sigma), path(\sigma') \in \mathcal{P}(CT(\mathcal{TN}))$ and $\star \in \{one, many\}$, $path(\sigma')$ is a $\star$-extension of $path(\sigma)$ if σ' is a $\star$-extension of σ.

Example 4 Consider the time Petri net $\widetilde{\mathcal{TN}}$ depicted in Fig. 4.1. For the firing schedule $\sigma = (t_1, 1)\,(t_4, 0)\,(t_3, 1)\,(t_2, 2)\,(t_5, 2) \in \mathcal{FS}(\widetilde{\mathcal{TN}})$, we obtain $Abs(path(\sigma)) = (a, 1, \emptyset)\,(a, 0, \emptyset)\,(b, 1, \{2\})\,(a, 2, \{1\})\,(c, 2, \{1, 3\})$, using the causal net $\widetilde{TN}$ shown in Fig. 4.2. An initial fragment of the causal tree of $\widetilde{\mathcal{TN}}$ is drawn in Fig. 4.3.

For time Petri nets, we establish some relationships between their correct causal processes and the labeled paths in their causal trees.

Proposition 4.2 [7] *Given time Petri nets $\mathcal{TN}$ and $\mathcal{TN}'$, with their causal trees $CT(\mathcal{TN})$ and $CT(\mathcal{TN}')$, respectively,*

(i) *whenever $\pi \in CP(\mathcal{TN})$ and $\pi' \in CP(\mathcal{TN}')$ with an isomorphism $f : \eta(TN_\pi) \to \eta(TN_{\pi'})$, then $Abs(path(FS_\pi(\rho))) = Abs'(path(FS_{\pi'}(f(\rho))))$, for all linearizations ρ of TN_π;*

(ii) *whenever $\sigma \in \mathcal{FS}(\mathcal{TN})$ and $\sigma' \in \mathcal{FS}(\mathcal{TN}')$ such that $Abs(path(\sigma)) = Abs'(path(\sigma'))$, then there is an isomorphism $f : \eta(TN_{\pi_\sigma}) \to \eta(TN_{\pi_{\sigma'}})$ such that $f(\rho_\sigma) = \rho_{\sigma'}$.*

4.3 Behavioral Equivalences of TPNs

4.3.1 Trace Equivalences

Trace equivalence, which is a traditional representative of the linear-time approach, compares the behavior of the systems modeled in terms of their languages—the sets of all possible executions of the systems.

---

[3] Notice that in $CT(\mathcal{TN})$, for any vertex $\sigma \in V$, there is a unique path starting from the root and finishing in σ.

Fig. 4.3 The causal tree of $\widetilde{\mathcal{TN}}$

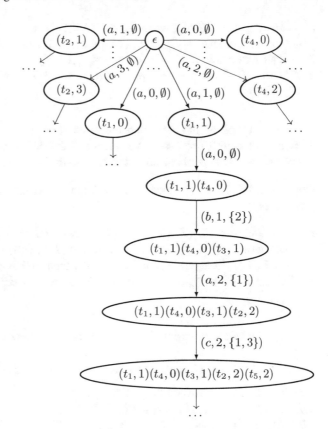

First, based on the TPN firing schedules labeled by actions with their time, we introduce the definition of the one-action and many-actions variants of interleaving trace equivalence.

Definition 4.5 Given (labeled over *Act*) time Petri nets $\mathcal{TN}$ and $\mathcal{TN}'$, and $\star = \{one, many\}$,

- – $Abs(\epsilon) \in \mathcal{L}_{int}^{\star}(\mathcal{TN})$, where $\epsilon \in \mathcal{FS}(\mathcal{TN})$;
  – whenever $Abs(\sigma) \in \mathcal{L}_{int}^{\star}(\mathcal{TN})$, where $\sigma \in \mathcal{FS}(\mathcal{TN})$, and $\sigma' \in \mathcal{FS}(\mathcal{TN})$ is a $\star$-extension of σ, then $Abs(\sigma') \in \mathcal{L}_{int}^{\star}(\mathcal{TN})$;
- $\mathcal{TN}$ and $\mathcal{TN}'$ are $_{int}^{\star}$-*trace equivalent* (denoted $\mathcal{TN} \equiv_{int}^{\star} \mathcal{TN}'$) iff $\mathcal{L}_{int}^{\star}(\mathcal{TN}) = \mathcal{L}_{int}^{\star}(\mathcal{TN}')$.

Secondly, we define trace equivalences built on the labeled paths and their extensions in the TPN causal trees, which, as far as we know, have not been considered in the Petri net literature.

Definition 4.6 Given (labeled over *Act*) time Petri nets $\mathcal{TN}$ and $\mathcal{TN}'$, and $\star = \{one, many\}$,

- – $Abs(u_0) \in \mathcal{L}_{ct}^{\star}(\mathcal{TN})$, where $u_0 \in \mathcal{P}(CT(\mathcal{TN}))$;
  – whenever $Abs(u) \in \mathcal{L}_{ct}^{\star}(\mathcal{TN})$, where $u \in \mathcal{P}(CT(\mathcal{TN}))$, and $u' \in \mathcal{P}(CT(\mathcal{TN}))$
    is a $\star$-extension of u, then $Abs(u') \in \mathcal{L}_{ct}^{\star}(\mathcal{TN})$;
- $\mathcal{TN}$ and $\mathcal{TN}'$ are $_{ct}^{\star}$-trace equivalent (denoted $\mathcal{TN} \equiv_{ct}^{\star} \mathcal{TN}'$) iff $\mathcal{L}_{ct}^{\star}(\mathcal{TN}) = \mathcal{L}_{ct}^{\star}(\mathcal{TN}')$.

Thirdly, based on the equivalence classes of the time posets obtained from the TPN correct causal processes, we formulate the definition of the one-action and many-actions variants of partial order trace equivalence.

Definition 4.7 Given (labeled over Act) time Petri nets $\mathcal{TN}$ and $\mathcal{TN}'$, and $\star = \{one, many\}$,

- – $[\eta(TN_0)]_{\simeq} \in \mathcal{L}_{po}^{\star}(\mathcal{TN})$, where $(TN_0, \varphi_0) \in \mathcal{CP}(\mathcal{TN})$;
  – whenever $[\eta(TN)]_{\simeq} \in \mathcal{L}_{po}^{\star}(\mathcal{TN})$, where $(TN, \varphi) \in \mathcal{CP}(\mathcal{TN})$, and (TN', φ')
    $\in \mathcal{CP}(\mathcal{TN})$ is a $\star$-extension of (TN, φ), then $[\eta(TN')]_{\simeq} \in \mathcal{L}_{po}^{\star}(\mathcal{TN})$.
- $\mathcal{TN}$ and $\mathcal{TN}'$ are $_{po}^{\star}$-trace equivalent (denoted $\mathcal{TN} \equiv_{po}^{\star} \mathcal{TN}'$) iff $\mathcal{L}_{po}^{\star}(\mathcal{TN}) = \mathcal{L}_{po}^{\star}(\mathcal{TN}')$.

We present alternative definitions of the trace equivalences more suitable for proof purposes.

Lemma 4.2 *Given a (labeled over Act) time Petri net $\mathcal{TN}$ and $\star = \{one, many\}$,*

(i) $\mathcal{L}_{int}^{\star}(\mathcal{TN}) = \{Abs(\sigma) \mid \sigma \in \mathcal{FS}(\mathcal{TN})\}$;
(ii) $\mathcal{L}_{ct}^{\star}(\mathcal{TN}) = \{Abs(u) \mid u \in \mathcal{P}(CT(\mathcal{TN}))\}$;
(iii) $\mathcal{L}_{po}^{\star}(\mathcal{TN}) = \{[\eta(TN)]_{\simeq} \mid (TN, \varphi) \in \mathcal{CP}(\mathcal{TN})\}$.

Proof Sketch (i) ((ii), (iii)) Straightforward by induction on the length of a firing schedule $\sigma \in \mathcal{FS}(\mathcal{TN})$ (a path $u \in \mathcal{P}(CT(\mathcal{TN}))$, a linearization ρ of TN in $(TN, \varphi) \in \mathcal{CP}(\mathcal{TN})$). $\qquad \square$

We are now ready to establish the interrelations between the trace equivalences considered above.

Proposition 4.3 *Given (labeled over Act) time Petri nets $\mathcal{TN}$ and $\mathcal{TN}'$, and $\bullet = \{int, ct, po\}$, $\star = \{one, many\}$,*

(i) $\mathcal{TN} \equiv_{ct}^{\star} \mathcal{TN}' \Rightarrow \mathcal{TN} \equiv_{int}^{\star} \mathcal{TN}'$;
(ii) $\mathcal{TN} \equiv_{ct}^{\star} \mathcal{TN}' \iff \mathcal{TN} \equiv_{po}^{\star} \mathcal{TN}'$;
(iii) $\mathcal{TN} \equiv_{\bullet}^{one} \mathcal{TN}' \iff \mathcal{TN} \equiv_{\bullet}^{many} \mathcal{TN}'$.

Proof (i) Directly follows from Definitions 4.4–4.6.

(ii) ($\Rightarrow$) Suppose $\mathcal{L}_{ct}^{\star}(\mathcal{TN}) = \mathcal{L}_{ct}^{\star}(\mathcal{TN}')$. Check that $\mathcal{L}_{po}^{\star}(\mathcal{TN}) = \mathcal{L}_{po}^{\star}(\mathcal{TN}')$. Consider an arbitrary element $[\eta(TN)]_{\simeq} \in \mathcal{L}_{po}^{\star}(\mathcal{TN})$, where $\pi = (TN, \varphi) \in \mathcal{CP}(\mathcal{TN})$. Due to Proposition 4.1(i), there is a unique firing schedule $\sigma_{\pi, \rho} = FS_{\pi}(\rho) \in \mathcal{FS}(\mathcal{TN})$, for any linearization ρ of TN. Consider an arbitrary linearization ρ of TN. Thanks to Definition 4.4 and Lemma 4.2(ii), we have that

$Abs(path(\sigma_{\pi,\rho})) \in \mathcal{L}^{\star}_{ct}(\mathcal{TN}) = \mathcal{L}^{\star}_{ct}(\mathcal{TN'})$. This means that there exists a path $path(\sigma') \in \mathcal{P}(CT(\mathcal{TN'}))$ such that $Abs'(path(\sigma')) = Abs(path(\sigma_{\pi,\rho}))$. By Proposition 4.1(i) and Lemma 4.2(iii), it holds that $[\eta(TN_{\pi_{\sigma'}})]_{\simeq} \in \mathcal{L}^{\star}_{po}(\mathcal{TN'})$. Moreover, according to Proposition 4.2(ii), we can find an isomorphism $f : \eta(TN_{\pi_{\sigma_{\pi,\rho}}}) \longrightarrow \eta(TN_{\pi_{\sigma'}})$ such that $f(\rho) = \rho_{\sigma'}$. Hence, $[\eta(TN_{\pi_{\sigma_{\pi,\rho}}})]_{\simeq} = [\eta(TN)]_{\simeq} \in \mathcal{L}^{\star}_{po}(\mathcal{TN'})$, due to Corollary 4.1(i).

($\Longleftarrow$) The proof follows similar lines of that in the opposite direction.

(iii) Straightforward using Lemma 4.2. □

4.3.2 Testing Equivalences

The concept of testing equivalence put forward in [15] is explicitly based on extracting information about the system behavior by testing it. Two processes are considered testing equivalent if, after each of the same executions, called experiments, the processes can continue with some action from the same sets of subsequent actions, called tests. As far as we know, the many-actions variant of tests has not been investigated in the setting of the equivalence.

Interleaving testing equivalence within TPNs deals with sequences of actions with their time (experiments) and sets of non-empty sequences of actions with their time (tests) checked after experiments. Here, both experiments and tests represent interleaving semantics. The length of the test sequences is equal to or greater than one in the case with many-actions extensions.

Definition 4.8 Given (labeled over Act) time Petri nets $\mathcal{TN}$ and $\mathcal{TN'}$, and $\star = \{one, many\}$,

- for a sequence $w \in (Act \times \mathbb{T})^*$ and a set $W \subseteq (Act \times \mathbb{T})^+$, $\mathcal{TN}$ **after** w **MUST**$^{\star}_{int}$ W iff for all firing schedules $\sigma \in \mathcal{FS}(\mathcal{TN})$ such that $Abs(\sigma) = w$, there exists an element $\hat{w} \in W$ and a firing schedule $\sigma' \in \mathcal{FS}(\mathcal{TN})$ such that σ' is a $\star$-extension of σ and $Abs(\sigma') = w\hat{w}$;
- $\mathcal{TN}$ and $\mathcal{TN'}$ are $^{\star}_{int}$-testing equivalent (denoted $\mathcal{TN} \approx^{\star}_{int} \mathcal{TN'}$) iff for all sequences $w \in (Act \times \mathbb{T})^*$ and for all sets $W \subseteq (Act \times \mathbb{T})^+$, it holds:

$$\mathcal{TN} \textbf{ after } w \textbf{ MUST}^{\star}_{int} W \iff \mathcal{TN'} \textbf{ after } w \textbf{ MUST}^{\star}_{int} W.$$

Next, we develop the definition of testing equivalence based on labeled paths in the causal trees of TPNs. In so doing, we consider experiments as sequences over the alphabet $(Act \times \mathbb{T} \times 2^{\mathbb{N}})$ (corresponding to labeled paths from the roots in the causal trees) and specify tests as sets of non-empty sequences over the same alphabet (corresponding to sets of extensions of the labeled paths in the causal trees). Here, both experiments and tests represent a combination of interleaving and partial order semantics. The length of the test sequences is equal to or greater than one in the case with many-actions extensions.

Definition 4.9 Given (labeled over Act) time Petri nets $\mathcal{TN}$ and $\mathcal{TN'}$ with their causal trees $CT(\mathcal{TN})$ and $CT(\mathcal{TN'})$, respectively, and $\star = \{one, many\}$,

- for a sequence $\omega \in (Act \times \mathbb{T} \times 2^{\mathbb{N}})^*$ and a set $\mathcal{W} \subseteq (Act \times \mathbb{T} \times 2^{\mathbb{N}})^+$, $\mathcal{TN}$ **after** ω **MUST**$^\star_{ct}$ $\mathcal{W}$ iff for all paths $u \in \mathcal{P}(\mathcal{CT}(\mathcal{TN}))$ such that $Abs(u) = \omega$, there exists an element $\widehat{\omega} \in \mathcal{W}$ and a path $u' \in \mathcal{P}(\mathcal{CT}(\mathcal{TN}))$ such that u' is a $\star$-extension of u and $Abs(u') = \omega\widehat{\omega}$;
- $\mathcal{TN}$ and $\mathcal{TN}'$ are causal tree testing equivalent ($\mathcal{TN} \approx^\star_{ct} \mathcal{TN}'$) iff for all sequences $\omega \in (Act \times \mathbb{T} \times 2^{\mathbb{N}})^*$ and sets $\mathcal{W} \subseteq (Act \times \mathbb{T} \times 2^{\mathbb{N}})^+$, it holds:

$$\mathcal{TN} \text{ after } \omega \text{ MUST}^\star_{ct} \mathcal{W} \iff \mathcal{TN}' \text{ after } \omega \text{ MUST}^\star_{ct} \mathcal{W}.$$

The idea of partial order testing is that experiments on the TPN are time posets and tests examined after experiments are sets of one-action or many-actions extensions of experiments. This contrasts with partial order based testing investigated in [7], where tests contain sets of time posets extending experiments by single actions with their time.

From now on, for the time poset $TP \in \mathcal{TP}_{Act}$ and $\star = \{one, many\}$, we use $\mathbb{TP}^\star_{TP} \subseteq \mathcal{TP}_{Act}$ to denote the set of $\star$-extensions of TP.

Definition 4.10 Given (labeled over Act) time Petri nets $\mathcal{TN}$ and $\mathcal{TN}'$, and $\star = \{one, many\}$,

- for a time poset $TP \in \mathcal{TP}_{Act}$ and a set $\mathbb{TP} \subseteq \mathbb{TP}^\star_{TP}$, $\mathcal{TN}$ **after** TP **MUST**$^\star_{po}$ $\mathbb{TP}$ iff for all causal processes $\pi = (TN, \varphi) \in \mathcal{CP}(\mathcal{TN})$ and for all isomorphisms $f : \eta(TN) \longrightarrow TP$, there exists a time poset $TP' \in \mathbb{TP}$, a causal process $\pi' = (TN', \varphi') \in \mathcal{CP}(\mathcal{TN})$, and an isomorphism $f' : \eta(TN') \longrightarrow TP'$, such that π' is a $\star$-extension of π and $f \subset f'$;
- $\mathcal{TN}$ and $\mathcal{TN}'$ are *poset testing equivalent* (denoted $\mathcal{TN} \approx^\star_{po} \mathcal{TN}'$) iff for all time posets $TP \in \mathcal{TP}_{Act}$ and for all sets $\mathbb{TP} \subseteq \mathbb{TP}^\star_{TP}$, it holds:

$$\mathcal{TN} \text{ after } TP \text{ MUST}^\star_{po} \mathbb{TP} \iff \mathcal{TN}' \text{ after } TP \text{ MUST}^\star_{po} \mathbb{TP}.$$

Proposition 4.4 *Given (labeled over Act) time Petri nets $\mathcal{TN}$ and $\mathcal{TN}'$, and $\bullet = \{int, ct, po\}$, $\star = \{one, many\}$,*

(i) $\mathcal{TN} \approx^\star_{ct} \mathcal{TN}' \Rightarrow \mathcal{TN} \approx^\star_{int} \mathcal{TN}'$;
(ii) $\mathcal{TN} \approx^\star_{ct} \mathcal{TN}' \iff \mathcal{TN} \approx^\star_{po} \mathcal{TN}'$.

Proof (i) Immediately follows from Definitions 4.8 and 4.9.

(ii) ($\Leftarrow$) Consider the proof of the case with $* = many$ (the proof of the case with $* = one$ is similar). Assume $\mathcal{TN} \approx^{many}_{po} \mathcal{TN}'$. Using Definition 4.10 and Lemma 4.2(iii), it is easy to see that $\mathcal{L}^{many}_{po}(\mathcal{TN}) = \mathcal{L}^{many}_{po}(\mathcal{TN}')$. By Proposition 4.3(ii), we have $\mathcal{L}^{many}_{ct}(\mathcal{TN}) = \mathcal{L}^{many}_{ct}(\mathcal{TN}')$. We shall check that $\mathcal{TN} \approx^{many}_{ct}$ $\mathcal{TN}'$. Take arbitrary $\omega \in (Act \times \mathbb{T} \times 2^{\mathbb{N}})^*$ and $\mathcal{W} \subseteq (Act \times \mathbb{T} \times 2^{\mathbb{N}})^+$. Suppose that $\mathcal{TN}$ **after** ω **MUST**$^{many}_{ct}$ $\mathcal{W}$. Then, we get $\mathcal{W} \neq \emptyset$.

W.l.o.g. assume $\omega = (a_1, \theta_1, K_1) \ldots (a_n, \theta_n, K_n)$ ($n \geq 0$). In the case with $\omega \notin \mathcal{L}^{many}_{ct}(\mathcal{TN}) = \mathcal{L}^{many}_{ct}(\mathcal{TN}')$, the result is obvious. Consider the case with $\omega \in \mathcal{L}^{many}_{ct}(\mathcal{TN}) = \mathcal{L}^{many}_{ct}(\mathcal{TN}')$. Then, we can find an arbitrary path $u_\omega = path(\sigma_\omega)$ in

$CT(TN)$ such that $Abs(u_\omega) = \omega$. According to Proposition 4.1(ii), there is a unique (up to an isomorphism) causal process $\pi_\omega = (TN_\omega, \varphi_\omega) \in CP(TN)$ with a unique linearization $\rho_\omega = e_1^\omega \ldots e_n^\omega$ of TN_ω such that $FS_{\pi_\omega}(\rho_\omega) = \sigma_\omega$. By Lemma 4.2(iii), we get $[\eta(TN_\omega)]_\simeq \in \mathcal{L}_{po}^{many}(TN)$. Take an arbitrary time poset $TP_\omega = (E_\omega, \preceq_\omega, Abs_\omega, \tau_\omega) \in [\eta(TN_\omega)]_\simeq$ with an arbitrary isomorphism $f_\omega : \eta(TN_\omega) \longrightarrow TP_\omega$. W.l.o.g. assume $f_\omega(e_j^\omega) = \widetilde{e}_j \in E_\omega$, for all $1 \le j \le n$.

For an arbitrary $\widehat{\omega} = (a_{n+1}, \theta_{n+1}, K_{n+1}) \ldots (a_{n+m}, \theta_{n+m}, K_{n+m}) \in \mathcal{W} \ne \emptyset$ ($m \ge 1$), construct the structure $TP_{\omega\widehat{\omega}} = TP_{n+m} = (E_{n+m}, \preceq_{n+m}, Abs_{n+m}, \tau_{n+m})$, by induction on $n \le j \le n + m$, as follows:

$j = n.$ $TP_j = TP_\omega.$

$j > n.$ Set $j = j + 1$; $E_j = E_{j-1} \cup \{\widetilde{e}_j\}$; $\preceq_j = \preceq_{j-1} \cup \{(\widetilde{e}_j, \widetilde{e}_j)\} \cup \{(\widetilde{e}_{j-r}, \widetilde{e}_j) \mid r \in K_j\}$; $Abs_j \mid_{E_{j-1}} = Abs_{j-1}$, $Abs_j(\widetilde{e}_j) = a_j$; $\tau_j \mid_{E_{j-1}} = \tau_{j-1}$, $\tau_j(\widetilde{e}_j) = \tau_{j-1}(\widetilde{e}_{j-1}) + \theta_j$.

It is routine to show that $TP_{\omega\widehat{\omega}}$ is a time poset, and, moreover, $TP_{\omega\widehat{\omega}}$ is a many-extension of TP_ω. Determine the set $\mathbb{TP}_\mathcal{W} = \{TP_{\omega\widehat{\omega}} \mid \widehat{\omega} \in \mathcal{W}\}$.

We shall show that TN **after** TP_ω $\textbf{MUST}_{po}^{many}$ $\mathbb{TP}_\mathcal{W}$. Take an arbitrary $\pi_\omega^1 = (TN_\omega^1, \varphi_\omega^1) \in CP(TN)$ and an arbitrary isomorphism $f_\omega^1 : \eta(TN_\omega^1) \longrightarrow TP_\omega$. As $[TP_\omega]_\simeq \in \mathcal{L}_{po}^{many}(TN)$, such π_ω^1 and f_ω^1 exist. Then, $f : \eta(TN_\omega) \longrightarrow \eta(TN_\omega^1) = (f_\omega^1)^{-1} \cdot f_\omega$ is an isomorphism. From Proposition 4.2(i), it follows that $Abs(u_\omega = path(\sigma_\omega = FS_{\pi_\omega}(\rho_\omega))) = \omega = Abs(u_\omega^1 = path(\sigma_\omega^1 = FS_{\pi_\omega^1}(\rho_\omega^1 = f(\rho_\omega))))$, where $\rho_\omega^1 = f(e_1^\omega \ldots e_n^\omega)$ is a linearization of TN_ω^1. As TN **after** ω $\textbf{MUST}_{ct}^{many}$ $\mathcal{W}$, for the path u_ω^1, we can find an element $\widehat{\omega} \in \mathcal{W}$ and a path $u_{\omega\widehat{\omega}}^1 = path(\sigma_{\omega\widehat{\omega}}^1) \in \mathcal{P}(CT(TN))$ such that $Abs(u_{\omega\widehat{\omega}}^1) = \omega\widehat{\omega}$ and $u_{\omega\widehat{\omega}}^1$ is a many-extension of u_ω^1, i.e. $\sigma_{\omega\widehat{\omega}}^1$ is a many-extension of σ_ω^1. Thanks to Lemma 4.1(ii), $\pi_{\omega\widehat{\omega}}^1 = \pi_{\sigma_{\omega\widehat{\omega}}^1} = (TN_{\omega\widehat{\omega}}^1, \varphi_{\omega\widehat{\omega}}^1)$ is a many-extension of $\pi_{\sigma_\omega^1}$ in TN and $\rho_{\omega\widehat{\omega}}^1 = \rho_{\sigma_{\omega\widehat{\omega}}^1} = \rho_{\sigma_\omega^1} e_{n+1}^1 \ldots e_{n+m}^1$. Due to Corollary 4.1(i), we can w.l.o.g. assume that $\pi_\omega^1 = \pi_{\sigma_\omega^1}$ and, hence, $\rho_\omega^1 = \rho_{\sigma_\omega^1}$. It is easy to specify an isomorphism $f_{\omega\widehat{\omega}}^1 : \eta(TN_{\omega\widehat{\omega}}^1) \longrightarrow TP_{\omega\widehat{\omega}} \in \mathbb{TP}_\mathcal{W}$ such that $f_\omega^1 \subset f_{\omega\widehat{\omega}}^1$. So, TN **after** TP_ω $\textbf{MUST}_{po}^{many}$ $\mathbb{TP}_\mathcal{W}$. Hence, by the assumption of the proposition, it holds that TN' **after** TP_ω $\textbf{MUST}_{po}^{many}$ $\mathbb{TP}_\mathcal{W}$. W.l.o.g. assume that for every $\widehat{\omega} \in \mathcal{W}$, we can find $\pi_{\omega\widehat{\omega}}^1 = (TN_{\omega\widehat{\omega}}^1, \varphi_{\omega\widehat{\omega}}^1) \in CP(TN)$ with a linearization $\rho_{\omega\widehat{\omega}}^1$ of $TN_{\omega\widehat{\omega}}^1$ such that $\sigma_{\omega\widehat{\omega}}^1 = FS_{\pi_{\omega\widehat{\omega}}^1}(\rho_{\omega\widehat{\omega}}^1)$, and an isomorphism $f_{\omega\widehat{\omega}}^1 : \eta(TN_{\omega\widehat{\omega}}^1) \longrightarrow TP_{\omega\widehat{\omega}}$, as specified above.

Verify that TN' **after** ω $\textbf{MUST}_{ct}^{many}$ $\mathcal{W}$. Take an arbitrary path $u_\omega^2 = path(\sigma_\omega^2) \in \mathcal{P}(CT(TN'))$ such that $Abs'(u_\omega^2) = \omega$. Since $\omega \in \mathcal{L}_{ct}^{many}(TN')$, there is at least one such path u_ω^2 in $CT(TN')$. According to Proposition 4.1(ii), we can find a unique (up to an isomorphism) causal process $\pi_\omega^2 = (TN_\omega^2, \varphi_\omega^2) \in CP(TN')$ with a unique linearization $\rho_\omega^2 = e_1^2 \ldots e_n^2$ of TN_ω^2 such that $FS_{\pi_\omega^2}(\rho_\omega^2) = \sigma_\omega^2$. Thanks to Proposition 4.2(ii), we get an isomorphism $f_\omega^2 : \eta(TN_\omega^2) \longrightarrow TP_\omega = (f_\omega : \eta(TN_\omega) \longrightarrow TP_\omega) \cdot (f' : \eta(TN_\omega^2) \longrightarrow \eta(TN_\omega))$ such that $f'(\rho_\omega^2) = \rho_\omega$. As TN' **after** TP_ω $\textbf{MUST}_{po}^{many}$ $\mathbb{TP}_\mathcal{W}$, we can find $TP_2' \in \mathbb{TP}_\mathcal{W}$, $\pi_2' = (TN_2', \varphi_2') \in CP(TN')$, and an isomorphism $f_2' : \eta(TN_2') \longrightarrow TP_2'$ such that π_2' is a many-extension of π_ω^2 and $f_\omega^2 \subset f_2'$. By the construction of $\mathbb{TP}_\mathcal{W}$, we can w.l.o.g. suppose $TP_2' = TP_{\omega\widehat{\omega}}$, for some $\widehat{\omega} \in \mathcal{W}$. Then, we have $\pi_{\omega\widehat{\omega}}^1 = (TN_{\omega\widehat{\omega}}^1, \varphi_{\omega\widehat{\omega}}^1) \in CP(TN)$ with a linearization

$\rho^1_{\omega\widehat{\omega}}$ of $TN^1_{\omega\widehat{\omega}}$ such that $\sigma^1_{\omega\widehat{\omega}} = FS_{\pi^1_{\omega\widehat{\omega}}}(\rho^1_{\omega\widehat{\omega}})$, and an isomorphism $f^1_{\omega\widehat{\omega}} : \eta(TN^1_{\omega\widehat{\omega}}) \longrightarrow TP_{\omega\widehat{\omega}}$, as specified above. Define the isomorphism $f'' : \eta(TN^1_{\omega\widehat{\omega}}) \longrightarrow \eta(TN'_2) = (f'_2)^{-1} \cdot f^1_{\omega\widehat{\omega}}$. From Proposition 4.2(i), it follows that $Abs(u^1_{\omega\widehat{\omega}} = path(\sigma^1_{\omega\widehat{\omega}} = FS_{\pi^1_{\omega\widehat{\omega}}}(\rho^1_{\omega\widehat{\omega}}))) = \omega\widehat{\omega} = Abs'(u^2_{\omega\widehat{\omega}} = path(\sigma^2_{\omega\widehat{\omega}} = FS_{\pi'_2}(\rho^2_{\omega\widehat{\omega}} = f''(\rho^1_{\omega\widehat{\omega}}))))$. Since $\rho^2_\omega = (f')^{-1}(\rho_\omega) = (f')^{-1} \cdot f^{-1}(\rho^1_\omega) = (f^2_\omega)^{-1} \cdot f^1_\omega(\rho^1_\omega)$, $\rho^2_{\omega\widehat{\omega}} = f''(\rho^1_{\omega\widehat{\omega}})$, and $(f^2_\omega)^{-1} \cdot f^1_\omega \subset (f^2_\omega)^{-1} \cdot f^1_{\omega\widehat{\omega}} = f''$, we get that $\sigma^2_{\omega\widehat{\omega}}$ is a many-extension of σ^2_ω, i.e. $u^2_{\omega\widehat{\omega}}$ is a many-extension of u^2_ω. Thus, due to an arbitrary choice of the path $u^2_\omega \in \mathcal{P}$ $(CT(TN'))$ such that $Abs'(u^2_\omega) = \omega$, it holds TN **after** ω **MUST**$^{many}_{ct}$ $W \Rightarrow TN'$ **after** ω **MUST**$^{many}_{ct}$ W. By symmetry, we obtain TN' **after** ω **MUST**$^{many}_{ct}$ $W \Rightarrow TN$ **after** ω **MUST**$^{many}_{ct}$ W.

(ii) ($\Rightarrow$) The proof follows similar lines of that in the opposite direction. $\qquad\square$

4.3.3 Bisimulation Equivalences

We start with introducing the one-action and many-actions variants of interleaving bisimulation between two TPNs, based on their firing schedules.

Definition 4.11 Given (labeled over Act) time Petri nets TN and TN', and $\star = \{one, many\}$, TN and TN' are $\underset{int}{\star}$-bisimilar (denoted $TN \underset{int}{\leftrightarrow^\star} TN'$) iff there is a $\underset{int}{\star}$-bisimilation $\alpha \subseteq FS(TN) \times FS(TN')$ between TN and TN', such that $(\epsilon, \epsilon) \in \alpha$, and, for all $(\sigma, \sigma') \in \alpha$, the following holds:

(a) $Abs(\sigma) = Abs'(\sigma')$;
(b) whenever σ_1 is a $\star$-extension of σ in TN, then there exists a pair $(\sigma_1, \sigma'_1) \in \alpha$ such that σ'_1 is a $\star$-extension of σ' in TN';
(c) similar to item (b) but the roles of TN and TN' swap round.

Secondly, we consider the definition of the two variants of causal tree bisimulation in the context of TPNs, which, to the best of our knowledge, have not been considered in the Petri net literature.

Definition 4.12 Given (labeled over Act) time Petri nets TN and TN', and $\star = \{one, many\}$, TN and TN' are $\underset{ct}{\star}$-bisimilar (denoted $TN \underset{ct}{\leftrightarrow^\star} TN'$) iff there is a $\underset{ct}{\star}$-bisimilation $\beta \subseteq \mathcal{P}(CT(TN)) \times \mathcal{P}(CT(TN'))$ between TN and TN', such that $(u_0, u'_0) \in \beta$, and, for all $(u, u') \in \beta$, the following holds:

(a) $Abs(u) = Abs'(path(u'))$;
(b) whenever u_1 is a $\star$-extension of u in $CT(TN)$, then there exists a pair $(u_1, u'_1) \in \beta$ such that u'_1 is a $\star$-extension of u' in $CT(TN')$;
(c) similar to item (b) but the roles of TN and TN' swap round.

Thirdly, we define the one-action and many-actions variants of partial order based bisimulation (in fact, history preserving bisimulation) between two TPNs, connecting their correct causal processes.

Definition 4.13 Given (labeled over Act) time Petri nets $\mathcal{TN}$ and $\mathcal{TN}'$, and $\star = \{one, many\}$, $\mathcal{TN}$ and $\mathcal{TN}'$ are ${}_{po}^{\star}$-bisimilar (denoted $\mathcal{TN} \underline{\leftrightarrow}_{po}^{\star} \mathcal{TN}'$) iff there is a ${}_{po}^{\star}$-bisimulation $\gamma \subseteq (\mathcal{CP}(\mathcal{TN}) \times \mathcal{CP}(\mathcal{TN}') \times \mathcal{F}(\mathcal{L}_{po}^{\star}(\mathcal{TN}), \mathcal{L}_{po}^{\star}(\mathcal{TN}')))^4$ between $\mathcal{TN}$ and $\mathcal{TN}'$, such that $(\pi_0, \pi_0', \emptyset) \in \gamma$, and, for all $(\pi, \pi', f) \in \gamma$, the following holds:

(a) $f : \eta(TN_\pi) \to \eta(TN_{\pi'})$ is an isomorphism;
(b) whenever π_1 is a $\star$-extension of π in $\mathcal{TN}$, then there exists a triple $(\pi_1, \pi_1', f_1) \in \gamma$ such that π_1' is a $\star$-extension of π' in $\mathcal{TN}'$ and $f \subset f_1$;
(c) similar to item (b) but the roles of $\mathcal{TN}$ and $\mathcal{TN}'$ swap round.

In [4], E. Best and others have shown, within the context of transition-labeled Petri nets, the coincidence of both variants of the extensions with both interleaving bisimulation and history preserving bisimulation. We extend their results to TPNs and establish a match between bisimulations in causal tree and partial order semantics.

Proposition 4.5 *Given (labeled over Act) time Petri nets $\mathcal{TN}$ and $\mathcal{TN}'$, and $\bullet = \{int, ct, po\}$, $\star = \{one, many\}$,*

(i) $\mathcal{TN} \underline{\leftrightarrow}_{ct}^{\star} \mathcal{TN}' \Rightarrow \mathcal{TN} \underline{\leftrightarrow}_{int}^{\star} \mathcal{TN}'$;
(ii) $\mathcal{TN} \underline{\leftrightarrow}_{ct}^{\star} \mathcal{TN}' \iff \mathcal{TN} \underline{\leftrightarrow}_{po}^{\star} \mathcal{TN}'$;
(iii) $\mathcal{TN} \underline{\leftrightarrow}_{\bullet}^{one} \mathcal{TN}' \iff \mathcal{TN} \underline{\leftrightarrow}_{\bullet}^{many} \mathcal{TN}'$.

Proof (i) Directly follows from Definitions 4.4, 4.11 and 4.12.

(ii) Suppose $\mathcal{TN} \underline{\leftrightarrow}_{ct}^{many} \mathcal{TN}'$, i.e. there exists a ${}_{ct}^{many}$-bisimulation $\beta \subseteq \mathcal{P}(CT(\mathcal{TN})) \times \mathcal{P}(CT(\mathcal{TN}'))$ between $\mathcal{TN}$ and $\mathcal{TN}'$. By Proposition 4.1(ii) and Proposition 4.2(ii), we get *Claim:* For any pair $(path(\sigma), path(\sigma')) \in \beta$, there is a triple $(\pi_\sigma, \pi_{\sigma'}, f_\sigma^{\sigma'})$.

Thanks to Claim, we can construct a relation $\gamma = \{(\pi_\sigma, \pi_{\sigma'}, f_\sigma^{\sigma'}) \mid (path(\sigma), path(\sigma')) \in \beta\}$ and check that γ is a ${}_{po}^{many}$-bisimulation between $\mathcal{TN}$ and $\mathcal{TN}'$. Clearly, $(\pi_0, \pi_0', \emptyset) \in \gamma$. Take an arbitrary triple $(\pi_\sigma, \pi_{\sigma'}, f_\sigma^{\sigma'}) \in \gamma$ and an arbitrary many-extension π_1 of π_σ in $\mathcal{TN}$. By virtue of Lemma 4.1(i), $\sigma\widehat{\sigma} = \sigma_{\pi_1,\rho_1}$ is a many-extension of $\sigma_{\pi_\sigma,\rho_\sigma}$ in $\mathcal{TN}$, for all linearizations $\rho_1 = \rho_\sigma \widehat{\rho}$ of TN_{π_1}. Consider an arbitrary $\rho_1 = \rho_\sigma \widehat{\rho}$ of TN_{π_1}. We can w.l.o.g. suppose that $\pi_1 = \pi_{\sigma\widehat{\sigma}}$, with $\rho_1 = \rho_{\sigma\widehat{\sigma}}$, by Corollary 4.1(i), and we get $\sigma_{\pi_\sigma,\rho_\sigma} = \sigma$, by Corollary 4.1(ii). Due to β being a ${}_{ct}^{many}$-bisimulation between $\mathcal{TN}$ and $\mathcal{TN}'$, there exists a pair $(path(\sigma\widehat{\sigma}), path(\sigma'\widehat{\sigma}')) \in \beta$ such that $\sigma'\widehat{\sigma}'$ is a many-extension of σ' in $\mathcal{TN}'$ and $Abs(path(\sigma\widehat{\sigma})) = Abs'(path(\sigma'\widehat{\sigma}'))$. Thanks to the construction of γ and Claim, we know that $(\pi_{\sigma\widehat{\sigma}}, \pi_{\sigma'\widehat{\sigma}'}, f_{\sigma\widehat{\sigma}}^{\sigma'\widehat{\sigma}'}) \in \gamma$. According to Lemma 4.1(ii), $\pi_{\sigma'\widehat{\sigma}'}$ is a many-extension of $\pi_{\sigma'}$ in $\mathcal{TN}'$. Obviously, $f_\sigma^{\sigma'} \subset f_{\sigma\widehat{\sigma}}^{\sigma'\widehat{\sigma}'}$. By symmetry, it is true that $\mathcal{TN} \underline{\leftrightarrow}_{po}^{many} \mathcal{TN}'$.

(iii) We consider the case with $\star = po$ (the other cases are similar).

($\Rightarrow$) Directly follows from Definition 4.13 and the deifinitions of the one- and many-extensions of the correct causal process of the TPN.

---

[4] $\mathcal{F}(\mathcal{L}_{po}^{*}(\mathcal{TN}), \mathcal{L}_{po}^{*}(\mathcal{TN}')) = \{f : TP \to TP' \mid f \text{ is a mapping}, [TP]_{\simeq} \in \mathcal{L}_{po}^{*}(\mathcal{TN}), [TP']_{\simeq} \in \mathcal{L}_{po}^{*}(\mathcal{TN}')\}$.

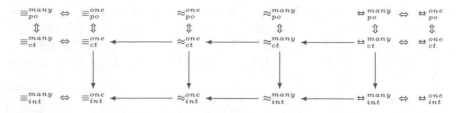

Fig. 4.4 The interrelations between the equivalences

($\Leftarrow$) Assume $\mathcal{TN} \underline{\leftrightarrow}_{po}^{one} \mathcal{TN}'$, i.e. there exists a $_{po}^{one}$-bisimulation $\gamma \subseteq \mathcal{CP}(\mathcal{TN}) \times \mathcal{CP}(\mathcal{TN}') \times \mathcal{F}(\mathcal{L}_{po}^{one}(\mathcal{TN}), \mathcal{L}_{po}^{one}(\mathcal{TN}'))$ between $\mathcal{TN}$ and $\mathcal{TN}'$. We shall show that γ is a $_{po}^{many}$-bisimulation between $\mathcal{TN}$ and $\mathcal{TN}'$. Take an arbitrary triple $(\pi, \pi', f) \in \gamma$ and an arbitrary many-extension π_1 of π in $\mathcal{TN}$. Clearly, the many-extension of π can be obtained by an iteration of one-extensions. Then, corresponding (related by γ) one-extensions in $\mathcal{TN}'$ iteratively obtained finally provide a many-extension of π' in $\mathcal{TN}'$ and a corresponding triple in γ satisfying item (b) with $\star = many$ of Definition 4.13. By symmetry, it holds that $\mathcal{TN} \underline{\leftrightarrow}_{po}^{many} \mathcal{TN}'$. □

4.4 Interrelations of the Equivalences

We start by establishing the interrelations between the equivalences under our consideration, in the framework of the whole class of TPNs.

Theorem 4.1 *Let* $\leftrightarrow, \rightleftharpoons \in \{\equiv, \approx, \underline{\leftrightarrow}\}$, $\star, \ast \in \{one, many\}$, *and* $\bullet, \circ \in \{int, ct, po\}$. *Given time Petri nets* $\mathcal{TN}$ *and* $\mathcal{TN}'$, $\mathcal{TN} \leftrightarrow_{\bullet}^{\star} \mathcal{TN}' \implies \mathcal{TN} \rightleftharpoons_{\circ}^{\ast} \mathcal{TN}'$ *iff there is a directed path from* $\leftrightarrow_{\bullet}^{\star}$ *to* $\rightleftharpoons_{\circ}^{\ast}$ *in Fig. 4.4.*

Proof ($\Leftarrow$) It is routine to show that all the implications (red arrows) in Fig. 4.4 follow from the Definitions and Propositions presented thus far.

($\Rightarrow$) We now demonstrate that it is impossible to draw any arrow from one equivalence to the other such that there is no directed path from the first equivalence to the second one in the graph in Fig. 4.4. For this purpose, we consider the time Petri nets depicted in Figs. 4.5, 4.6 and 4.7.

The time Petri nets $\mathcal{TN}_1$ and $\mathcal{TN}_2$ shown in Fig. 4.5 are $\underline{\leftrightarrow}_{int}^{many}$–equivalent but not $\equiv_{po}^{one}$–equivalent. Verify that the latter is true. Define a poset $TP = (\{x_1, x_2\}, \preceq, Abs, \tau)$ with $\preceq = \{(x_1, x_1), (x_2, x_2)\}$, $Abs(x_1) = a$, $Abs(x_2) = b$, $\tau(x_1) = \tau(x_2) = 1$. It is easy to see that there is a causal process $\pi_2 = (TN_2, \varphi_2) \in \mathcal{CP}(\mathcal{TN}_2)$ with E_{TN_2} containing two concurrent events labeled by a and b, both with time value 1, and an isomorphism $f_2: \eta(TN_2) \longrightarrow TP$. Then, we obtain $TP \in \mathcal{L}_{po}^{one}(\mathcal{TN}_2)$. However, for any causal process $\pi_1 = (TN_1, \varphi_1) \in \mathcal{CP}(\mathcal{TN}_1)$, there is no isomorphism $f_1: \eta(TN_1) \longrightarrow TP$. So, we have that $TP \notin \mathcal{L}_{po}^{one}(\mathcal{TN}_1)$.

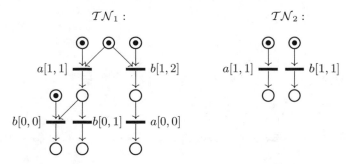

Fig. 4.5 $\leftrightarrow^{many}_{int}$–equivalent and non-$\equiv^{one}_{po}$–equivalent time Petri nets

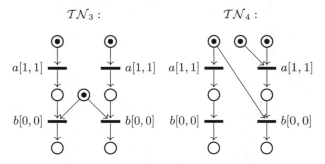

Fig. 4.6 $\equiv^{one}_{po}$–equivalent and non-$\approx^{one}_{int}$–equivalent time Petri nets

The time Petri nets $\mathcal{TN}_3$ and $\mathcal{TN}_4$ depicted in Fig. 4.6 are $\equiv^{one}_{po}$–equivalent. Check that they are not $\approx^{one}_{int}$–equivalent. For all firing schedules $\sigma \in \mathcal{FS}(\mathcal{TN}_3)$ such that $Abs_3(\sigma) = (a, 1)(b, 0)$, there exists an element $\widehat{w} = (a, 0) \in Act \times \mathbb{T}$ and a firing schedule $\sigma' \in \mathcal{FS}(\mathcal{TN}_3)$ such that σ' is a one-extension of σ and $Abs_3(\sigma') = (a, 1)(b, 0)(a, 0)$. However, this is not the case in $\mathcal{TN}_4$.

The time Petri nets $\mathcal{TN}_5$ and $\mathcal{TN}_6$ shown in Fig. 4.7 are $\approx^{one}_{po}$–equivalent but not $\approx^{many}_{int}$–equivalent. Let's make sure that the latter is true. For all firing schedules $\sigma \in \mathcal{FS}(\mathcal{TN}_6)$ such that $Abs_6(\sigma) = (a, 0)$, there exists an element $\widehat{w} = (b, 2.1)(b, 0.3) \in (Act \times \mathbb{T})^+$ and a firing schedule $\sigma' \in \mathcal{FS}(\mathcal{TN}_6)$ such that σ' is a many-extension of σ and $Abs_6(\sigma') = (a, 0)(b, 2.1)(b, 0.3)$. However, this is not the case in $\mathcal{TN}_5$.

The time Petri nets $\mathcal{TN}_7$ and $\mathcal{TN}_8$ depicted in Fig. 4.8 are $\approx^{many}_{po}$–equivalent but not $\leftrightarrow^{one}_{int}$–equivalent. Let us check the validity of the latter assertion. For a firing schedule $\sigma \in \mathcal{FS}(\mathcal{TN}_8)$ with $Abs_8(\sigma) = (a, 1)$, there exist two *one*-extensions of σ—σ_1 with $Abs_8(\sigma_1) = (a, 1)(b, 1)$ and σ_2 with $Abs_8(\sigma_2) = (a, 1)(c, 1)$. However, we cannot find any firing schedule in $\mathcal{FS}(\mathcal{TN}_7)$ to associate it with σ.   □

Next, we introduce the definitions of subclasses of TPNs. The TPN $\mathcal{TN} = (\mathcal{N}, D)$ is called

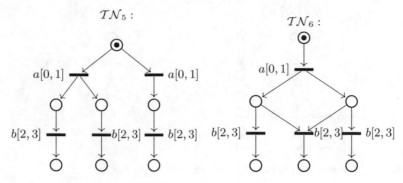

Fig. 4.7 $\approx_{po}^{one}$–equivalent and non-$\approx_{int}^{many}$–equivalent time Petri nets

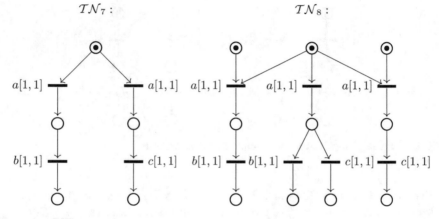

Fig. 4.8 $\approx_{po}^{many}$–equivalent and non-$\underleftrightarrow{}_{int}^{one}$–equivalent time Petri nets

- *deterministic* iff for any state $S = (M, I) \in RS(\mathcal{TN})$ and for any transitions $t, t' \in T$ ready to fire from S after some delay time $\theta \in \mathbb{T}$, if $Abs(t) = Abs(t')$, then $t = t'$;
- *sequential* iff for any state $S = (M, I) \in RS(\mathcal{TN})$ and for any transitions $t, t' \in T$ enabled at a marking M, $^{\bullet}t \cap {}^{\bullet}t' \neq \emptyset$;
- *zeroth* (untime) iff $D(t) = [0, 0]$, for all $t \in T$.

Consider characteristic properties of deterministic TPNs.

Lemma 4.3 *Given a deterministic time Petri net* $\mathcal{TN}$,

(i) *if* $Abs(\sigma_1) = Abs(\sigma_2)$, *then* $\sigma_1 = \sigma_2$, *for any* $\sigma_1, \sigma_2 \in \mathcal{FS}(\mathcal{TN})$;
(ii) *if* $Abs(u_1) = Abs(u_2)$, *then* $u_1 = u_2$, *for any* $u_1, u_2 \in \mathcal{P}(\mathcal{CT}(\mathcal{TN}))$;
(iii) *if* $\eta(TN_{\pi_1}) \simeq \eta(TN_{\pi_2})$, *then* $\pi_1 \simeq \pi_2$, *for any* $\pi_1, \pi_2 \in \mathcal{CP}(\mathcal{TN})$.

Proof Sketch (i) Directly follows from the definition of the deterministic TPN.
    (ii) Straightforward using Definition 4.4 and item (i).

(iii) Assume $f : \eta(TN_{\pi_1}) \simeq \eta(TN_{\pi_2})$ is an isomorphism. Take an arbitrary linearization ρ of TN_{π_1}. Clearly, $f(\rho)$ is a linearization of TN_{π_2}. Thanks to Proposition 4.2(i), there is a unique firing schedule $\sigma_1 = FS_{\pi_1}(\rho)$ $(\sigma_2 = FS_{\pi_2}(f(\rho))$ in $\mathcal{FS}(TN)$. Then, $Abs(\sigma_1) = Abs(\sigma_2)$ because f is an isomorphism. By item (i), we get $\varphi_{\pi_1}(e) = \varphi_{\pi_2}(f(e))$, for all $e \in E_{TN_{\pi_1}}$, i.e. $\pi_1 \simeq \pi_2$. $\square$

We demonstrate that the linear time—branching time spectrum collapses for deterministic TPNs.

Proposition 4.6 *Given deterministic time Petri nets TN and TN', and $\bullet \in \{int,$ $ct, po\}$,*

$$TN \equiv_{\bullet}^{one} TN' \iff TN \underleftrightarrow{}_{\bullet}^{many} TN'.$$

Proof Thanks to Theorem 4.1, it is sufficient to show the truth of statement in the direction $\Rightarrow$. Consider the proof for $\bullet = po$ (the proofs for the other cases are similar). Suppose $TN \equiv_{po}^{one} TN'$, i.e. $\mathcal{L}_{po}^{one}(TN) = \mathcal{L}_{po}^{one}(TN')$. Notice that for any $\pi \in \mathcal{CP}(TN)$ and $\pi' \in \mathcal{CP}(TN')$, there is at most one isomorphism $f : \eta(TN_\pi) \to \eta(TN_{\pi'})$. Construct a relation $\gamma = \{(\pi, \pi', f) | [\eta(TN_\pi)]_\sim = [\eta(TN_{\pi'})]_\sim \in \mathcal{L}_{po}^{one}(TN), f : \eta(TN_\pi) \to \eta(TN_{\pi'})$ is an isomorphism$\}$. We first check that γ is a $_{po}^{one}$-bisimulation between TN and TN'. Clearly, $(\pi_0, \pi_0', \emptyset) \in \gamma$. Consider an arbitrary triple $(\pi, \pi', f) \in \gamma$ and an arbitrary one-extension $\pi_1 \in \mathcal{CP}(TN)$ of π. Then, due to Lemma 4.2(iii), $[\eta(TN_{\pi_1})]_\simeq \in \mathcal{L}_{po}^{one}(TN) = \mathcal{L}_{po}^{one}(TN')$. Then, there is a causal process $\widetilde{\pi}_1' \in \mathcal{CP}(TN')$ such that $\eta(TN_{\widetilde{\pi}_1'}) \in [\eta(TN_{\pi_1})]_\simeq$. Let $\widetilde{f}_1 : \eta(TN_{\pi_1}) \longrightarrow \eta(TN_{\widetilde{\pi}_1'})$ be an isomorphism. Moreover, as π_1 is a one-extension of π, we can find $\widetilde{\pi}' \in \mathcal{CP}(TN')$ such that $\widetilde{\pi}_1'$ is a one-extension of $\widetilde{\pi}'$ with an isomorphism $\widetilde{f} : \eta(TN_\pi) \to \eta(TN_{\widetilde{\pi}'})$, where $\widetilde{f} = \widetilde{f}_1|_{E_{\eta(TN_\pi)}}$. Then, $f' : \eta(TN_{\pi'}) \to \eta(TN_{\widetilde{\pi}'}) = \widetilde{f} \cdot f^{-1}$ is an isomorphism. Suppose that there is no one-extension $\pi_1' \in \mathcal{CP}(TN')$ of π' such that $\eta(TN_{\pi_1'}) \simeq \eta(TN_{\widetilde{\pi}_1'})$. Thanks to Definition 4.3 and the definition of a one-extension of the causal process of the TPN, we get that $\varphi_{\pi'}(e) \neq \varphi_{\widetilde{\pi}'}(f'(e))$, for some $e \in E_{TN_{\pi'}}$, contradicting Lemma 4.3(iii) because $\eta(TN_{\pi'}) \simeq \eta(TN_{\widetilde{\pi}'})$. Then, it is true that $[\eta(TN_{\pi_1})]_\sim = [\eta(TN_{\pi_1'})]_\sim \in \mathcal{L}_{po}^{one}(TN')$. Let $f_1 : \eta(TN_{\pi_1}) \to \eta(TN_{\pi_1'})$ be an isomorphism. Hence, we have $(\pi_1, \pi_1', f_1) \in \gamma$, by the construction of γ. Clearly, $f \subset f_1$. By symmetry, it is true that $TN \underleftrightarrow{}_{po}^{one} TN'$. Thanks to Proposition 4.5(iii), it holds that $TN \underleftrightarrow{}_{po}^{many} TN'$.

$\square$

Consider characteristic features of sequential TPNs.

Lemma 4.4 *Given a sequential time Petri net TN,*

(i) *for any $\pi = (TN, \varphi) \in \mathcal{CP}(TN)$, $e_1 \prec_{TN} e_2 \ldots e_{n-1} \prec_{TN} e_n$, for a unique linearization $\rho = e_1 \ldots e_n$ of TN $(n \geq 0)$;*
(ii) *for any $\sigma = (t_1, \theta_1) \ldots (t_n, \theta_n) \in V_{\mathcal{CT}(TN)}$, $path(\sigma) = (t_1, \theta_1, K_1 = \emptyset) \ldots (t_n, \theta_n, K_n = \{1, \ldots, n-1\})$ $(n \geq 0)$.*

Proof Sketch (i) Consider arbitrary causal process $\pi = (TN, \varphi) \in \mathcal{CP}(TN)$ and linearization $\rho = e_1 \ldots e_n$ of TN $(n > 0)$. Suppose a contrary, i.e. for some elements

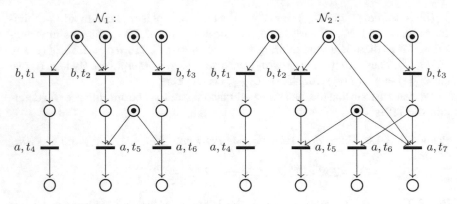

Fig. 4.9 $\leftrightarrow_{po}^{many}$–equivalent zeroth time Petri nets

$e \neq e'$ it holds that $e \smile e'$. Hence, $^{\bullet}\varphi(e) \cap {}^{\bullet}\varphi(e') = \emptyset$. So, using the definition of the homomorphism, we get a contradiction to the definition of the sequential TPN.

(ii) Immediately follows from Definition 4.4, Proposition 4.1(ii) and item (i). $\square$

We show that the interleaving—partial spectrum collapses for sequential TPNs.

Proposition 4.7 *Let $\leftrightarrow \in \{\equiv, \approx, \leftrightarrow\}$ and $\star \in \{one, many\}$. Given sequential time Petri nets $\mathcal{TN}$ and $\mathcal{TN}'$,*

$$\mathcal{TN} \leftrightarrow_{int}^{\star} \mathcal{TN}' \iff \mathcal{TN} \leftrightarrow_{po}^{\star} \mathcal{TN}'.$$

Proof Due to Propositions 4.3(ii)–4.5(ii), it is sufficient to check that $\mathcal{TN} \leftrightarrow_{int}^{\star} \mathcal{TN}' \iff \mathcal{TN} \leftrightarrow_{ct}^{\star} \mathcal{TN}'$. Thanks to Theorem 4.1, it is reasonable to show the truth of the statement only in the direction $\Rightarrow$. Consider the case with $\leftrightarrow = \leftrightarrow$ and $\star = many$ (the other cases are similar). Assume $\mathcal{TN} \leftrightarrow_{int}^{many} \mathcal{TN}'$. Then, there exists a $_{int}^{many}$-bisimulation $\alpha \subseteq \mathcal{FS}(\mathcal{TN}) \times \mathcal{FS}(\mathcal{TN}')$ between $\mathcal{TN}$ and $\mathcal{TN}'$. Construct $\beta = \{(path(\sigma), path(\sigma')) \mid (\sigma, \sigma') \in \alpha\} \subseteq \mathcal{P}(\mathcal{CT}(\mathcal{TN})) \times \mathcal{P}(\mathcal{CT}(\mathcal{TN}'))$. The result directly follows from Definitions 4.11, 4.12 and Lemma 4.4(ii). $\square$

Finally, we exemplify that the equivalences under consideration are incomparable for time Petri nets and zeroth (untime) Petri nets.

Example 5 The zeroth time Petri nets $\widetilde{\mathcal{TN}}_1 = (\mathcal{N}_1, D_1 = [0; 0])$ and $\widetilde{\mathcal{TN}}_2 = (\mathcal{N}_2, D_2 = [0; 0])$, where $\mathcal{N}_1$ and $\mathcal{N}_2$ are shown in Fig. 4.9, are $\leftrightarrow_{po}^{many}$–equivalent, whereas $\widetilde{\mathcal{TN}}_1' = (\mathcal{N}_1, D_1')$, with $D_1'(t) = [1, 1]$, for all $t \neq t_4 \in T_{\mathcal{N}_1}$, $D_1'(t_4) = [1, 4]$, and $\widetilde{\mathcal{TN}}_2' = (\mathcal{N}_2, D_2')$, with $D_2'(t) = [1, 1]$, for all $t \neq t_7 \in T_{\mathcal{N}_2}$, $D_2'(t_7) = [1, 4]$, are not $\equiv_{int}^{one}$–equivalent. The latter is true because there is the firing schedule $\sigma \in \mathcal{FS}(\widetilde{\mathcal{TN}}_1')$ such that $\widetilde{Abs}_1(\sigma) = (b, 1)(b, 1)(a, 3.5) \in \mathcal{L}_{int}^{one}(\widetilde{\mathcal{TN}}_1')$ but this is not the case in $\widetilde{\mathcal{TN}}_2'$.

On the other hand, the time Petri nets $\widetilde{\mathcal{TN}}_3 = (\mathcal{N}_3, D_3' = [1; 1])$ and $\widetilde{\mathcal{TN}}_4 = (\mathcal{N}_4, D_4' = [1; 1])$, where $\mathcal{N}_3$ and $\mathcal{N}_4$ are depicted in Fig. 4.10, are $\leftrightarrow_{po}^{many}$–equivalent,

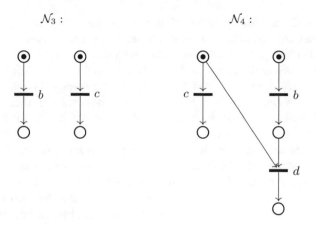

Fig. 4.10   Non-$\equiv_{int}^{one}$–equivalent zeroth time Petri nets

while their zeroth variants $\widetilde{\mathcal{TN}}_3 = (\mathcal{N}_3, D_3 = [0; 0])$ and $\widetilde{\mathcal{TN}}_4 = (\mathcal{N}_4, D_4 = [0; 0])$ are not $\equiv_{int}^{one}$–equivalent. The latter holds since there exists the firing schedule $\sigma \in \mathcal{FS}(\widetilde{\mathcal{TN}}_4)$ such that $\widetilde{Abs}_4(\sigma) = (b, 0)(d, 0) \in \mathcal{L}_{int}^{one}(\widetilde{\mathcal{TN}}_4)$ but this is not the case in $\widetilde{\mathcal{TN}}_3$.

4.5   Concluding Remarks

In this paper, we sought to provide a flexible abstract mechanism based on behavioral equivalences of linear time—branching time and interleaving—partial order spectra in order to consider time Petri nets as a basis for various approaches to describing and studying the behavior of concurrent and real time systems. For this aim, it is important to be able to use minimal TPNs equivalent nevertheless to the original larger ones, which guarantees the preservation of behavioral properties.

In particular, we have defined and compared the one-action and many-actions extensions in trace, testing, and bisimulation equivalences in various semantics in the setting of safe TPNs. In so doing, we have dealt with three representations of the TPN behavior: firing schedules presenting interleaving semantics; causal processes with partial orders derived from causal nets; and causal tree semantics constructed from the firing schedules and partial orders. It has turned out that the one-action and many-actions extensions coincide on trace and bisimulation equivalences in the semantic representations under consideration but this is not the case for testing equivalences. Within TPNs, we used close relationships, firstly, between their firing schedules and their correct causal processes, and, secondly, between the labeled paths in their causal trees and their correct causal processes. As the main result, the hierarchies of interrelations of the equivalences have been constructed for the whole class of TPNs and their various subclasses. The results obtained show, on the one hand, the discriminating power of the approaches of the two spectra and, on the other hand, the coincidence of the semantics based on causal trees and partial orders in a specific approach. When dealing with some particular subclasses of the model under consideration, we have demonstrated that there is no difference between more concrete and more abstract observations. Furthermore, although the semantics considered have turned out to be incomparable with the semantics on concrete untime Petri nets. However, the equivalence hierarchies developed here work for the untime case as well.

As for future work and modifications, we plan to see how the semantics and equivalences under consideration interrelate in the context of safe/bounded Petri nets with weak timing (no transition is ever forced to be fired) [6]. Another interesting line of further research should be extending the results obtained with invisible actions and with the one- and many-extensions of other equivalences known from the literature. Also, research is expected on a more precise understanding of the properties of systems preserved by different equivalences.

References

1. Andreeva, M., Virbitskaite, I.: Observational equivalences for timed stable event structures. Fundam. Inf. **72**(1–3), 1–19 (2006)
2. Aura, T., Lilius, J.: A causal semantics for time Petri nets. Theor. Comput. Sci. **243**(1–2), 409–447 (2000)
3. Bérard, B., Cassez, F., Haddad, S., Lime, D., Roux, O.H.: Comparison of the expressiveness of timed automata and time Petri nets. In: Pettersson, P., Yi, W. (eds.) Formal Modeling and Analysis of Timed Systems. FORMATS 2005. Lecture Notes in Computer Science, vol. 3829, pp. 211–225. Springer, Berlin (2005). https://doi.org/10.1007/11603009_17
4. Best, E., Devillers, R., Kiehn, A., Pomello, L.: Concurrent bisimulations in Petri Nets. Acta Inf. **28**(3), 231–264 (1991)
5. Bihler, E., Vogler, W.: Timed Petri nets: efficiency of asynchronous systems. In: Bernardo, M., Corradini, F. (eds.) Formal Methods for the Design of Real-Time Systems. SFM-RT 2004. Lecture Notes in Computer Science, vol. 3185, pp. 25–58. Springer, Berlin (2004). https://doi.org/10.1007/978-3-540-30080-9_2
6. Boyer, M., Roux, O.H.: On the compared expressiveness of arc, place and transition time Petri nets. Fundam. Inf. **88**(3), 225–249 (2008)
7. Bozhenkova, E., Virbitskaite, I.: Testing equivalences of time Petri nets. Program Comput. Softw. **46**(4), 251–260 (2020)
8. Bushin, D., Virbitskaite, I.: Comparative trace semantics of time Petri nets. Program Comput. Softw. **41**(3), 131–139 (2015)
9. Darondeau, Ph., Degano, P.: Refinement of actions in event structures and causal trees. Theor. Comput. Sci. **118**(1), 21–48 (1993)
10. van Glabbeek, R.J.: The linear time—branching time spectrum I: the semantics of concrete, sequential processes. In: Bergstra, J.A., Ponse, A., Smolka, S.A. (eds.) Handbook of Process Algebra, pp. 3–99. Elsevier (2001)
11. van Glabbeek, R.J., Goltz, U., Schicke, J.-W.: On causal semantics of Petri nets. In: Katoen, JP., König, B. (eds.) CONCUR 2011—Concurrency Theory. CONCUR 2011. Lecture Notes in Computer Science, vol. 6901, pp. 43–59. Springer, Berlin (2011). https://doi.org/10.1007/978-3-642-23217-6_4

12. Goltz, U., Wehrheim, H.: Causal testing. In: Penczek, W., Szalas, A. (eds.) Mathematical Foundations of Computer Science. MFCS 1996. Lecture Notes in Computer Science, vol. 1113, pp. 394–406. Springer, Berlin (1996). https://doi.org/10.1007/3-540-61550-4_165
13. Merlin, P., Faber, D.J.: Recoverability of communication protocols-implications of a theoretical study. IEEE Trans. Commun. **24**(9), 1036–1043 (1976)
14. De Nicola, R.: Behavioral equivalences. In: Padua, D. (ed.) Encyclopedia of Parallel Computing, pp. 120–127. Springer, Boston, MA (2011)
15. De Nicola, R., Hennessy, M.: Testing equivalence for processes. Theor. Comput. Sci. **34**, 83–133 (1984)
16. Nielsen, M., Rozenberg, G., Thiagarajan, P.S.: Behavioural notions for elementary net systems. Distrib. Comput. **4**(1), 45–57 (1990)
17. Pomello, L., Rozenberg, G., Simone, C.: A Survey of equivalence notions for net based systems. In: Rozenberg, G. (ed.) Advances in Petri Nets. Lecture Notes in Computer Science, vol. 609, pp. 410–472. Springer, Berlin (1992). https://doi.org/10.1007/3-540-55610-9_180
18. Rabinovich, A., Trakhtenbrot, B.A.: Behavior structures and nets. Fundam. Inf. **11**, 357–404 (1988)
19. Tarasyuk, I.V.: Equivalences for behavioural analysis of concurrent and distributed computing systems. "Geo" Publisher, Novosibirsk (2007) (in Russian)
20. Vogler, W.: Modular construction and partial order semantics of Petri nets. Lecture Notes in Computer Science, vol. 625. Springer, Berlin (1992)
21. Virbitskaite, I., Bushin, D., Best, E.: True concurrent equivalences in time Petri net. Fundam. Inf. **149**(4), 401–418 (2016)

Chapter 5
Toward Recommender Systems Scalability and Efficacy

Eyad Kannout, **Marek Grzegorowski**, and **Hung Son Nguyen**

Abstract Recommender systems play a key role in many branches of the digital economy. Their primary function is to select the most relevant services or products to users' preferences. The article presents selected recommender algorithms and their most popular taxonomy. We review the evaluation techniques and the most important challenges and limitations of the discussed methods. We also introduce Factorization Machines and Association Rules-based recommender system (FMAR) that addresses the problem of efficiency in generating recommendations while maintaining quality.

Keywords Recommendation systems · Collaborative filtering · Memory-based techniques · Model-based techniques · Content-based filtering · Hybrid filtering

5.1 Introduction

Along with the outspread of online services, social media, and eCommerce, many companies have highlighted a vital challenge to leveraging the data to boost their profits and improve their services and users' satisfaction. Recommender systems (RS) have become an established element in the digital world with many areas of their successful applications [29]. Especially after the COVID-19 pandemic, since online services have become a daily routine for most people. Therefore, the demand for finding more efficient techniques to generate recommendations have been increased more than ever before.

Recommender systems predict the utility of an item to a user concerning the user's preferences, and items may represent any services or things like movies, shops,

E. Kannout · M. Grzegorowski · H. Son Nguyen (✉)
Institute of Informatics, University of Warsaw, ul. Banacha 2, 02-097 Warsaw, Poland
e-mail: son@mimuw.edu.pl

E. Kannout
e-mail: eyad.kannout@mimuw.edu.pl

M. Grzegorowski
e-mail: m.grzegorowski@mimuw.edu.pl

© The Author(s), under exclusive license to Springer Nature Switzerland AG 2023
B. Schlingloff et al. (eds.), *Concurrency, Specification and Programming, Studies in Computational Intelligence 1091*, https://doi.org/10.1007/978-3-031-26651-5_5

events, etc. [30]. However, the particular expectations may differ between the service provider and the user of the RS [53]. Service providers typically aim to increase the number of sold products, diversify the displayed offer, increase user satisfaction, or better understand users' needs. On the other hand, users may desire to find new or surprising items, find a sequence or group of articles that fit well together or find suitable items given an existing context.

Technically, the task of recommender systems can be classified into prediction and recommendation [29]. In the first one, the RS estimates the rating of a particular item a particular user would assign, answering whether a user would like this specific item or not. In the recommendation task, the goal is to provide a list of N items that the user would be interested in (i.e., top-N recommendation problem). There are also many approaches to evaluating their quality depending on the investigated task assessing efficiency, scalability, accuracy, unexpectedness, serendipity, or diversity of the evaluated system [55].

Over the years, researchers have suggested various approaches for building recommender systems that utilize the rating history and possibly some other information, such as users' demographics and items' characteristics. Literature provides several taxonomies for RS [11]. The most common ones split RSs into three main categories:

- Collaborative filtering.
- Content-based filtering.
- Hybrid filtering

Collaborative Filtering (CF) is one of the most widely used and successful techniques, with excellent results in a wide range of applications in many fields [11]. Despite the noticeable decline in their popularity in favor of collaborative systems, content-based filtering (CBF) techniques are still widely used, primarily because they are well handling the so-called cold-start problem [44]. Due to significantly different characteristics of content and collaborative-based techniques, it is often advisable to construct various combinations to leverage their capabilities, which approach is referred to as hybrid filtering [2, 61].

In this chapter, we present a literature review pertaining to state-of-the-art recommendation systems, which is an extension of [40]. We also propose a Factorization Machines and Association Rules-based recommender system (FMAR). FMAR introduces a novel approach to generating association rules using several frequent pattern mining techniques. The main problem addressed by FMAR is to minimize the prediction latency of the recommender system, ensuring no significant impact on the recommendations' quality. Practitioners and businesses highly appreciate such a technique since the recommendations become useless if not produced in an acceptable time frame.

The remainder of this chapter is structured as follows. In Sect. 5.2, we provide a comprehensive overview of recommender systems methods and literature. In particular, Sect. 5.2.1 reviews the collaborative filtering approach including memory-based techniques and model-based techniques. Section 5.2.2 provides the rudimentary knowledge related to content-based filtering. In Sect. 5.2.3, we discuss the challenges and Limitations of both CF and CBF that propelled the emergence of hybrid methods—presented in Sect. 5.2.4. Section 5.2.5 surveys the most popular evaluation techniques of recommendation systems. In Sect. 5.3, we introduce a novel recommender system that utilizes Factorization Machines and Association Rules (FMAR) to estimate the user's ratings for new items. Section 5.4 evaluates the performance of the FMAR by comparing it with a traditional recommender system built without employing association rules. Finally, in Sect. 5.5, we summarise the chapter.

5.2 Rudiments of Recommender Systems

5.2.1 Collaborative Filtering

The basic idea behind collaborative filtering (CF) is that users who have historically similar tastes tend to behave similarly in the future. has proved to be very successful by both practitioners and researchers, and in both information filtering applications and E-commerce applications [30]. Basically, CF-based methods rely only on historical rating transactions to generate recommendations, meaning that the more ratings the users provide, the more accurate the recommendations become. The CF-based methods are usually distinguished by whether they operate over explicit ratings, i.e., when the user rates particular items, or implicit ones, i.e., when the ratings are inferred from observable user activity, such as products purchases, page views, mouse clicks, or any other types of information access patterns [30]. In the literature, collaborative filtering methods are classified into two main categories:

- Memory-based techniques.
- Model-based techniques.

In the following sections, we provide a detailed review of these methods.

5.2.1.1 Memory-Based Techniques

The memory-based techniques directly use the collected rating history to predict the rating of items that the user has not seen before. They rely on the stored users to items relation, typically represented as a user-item rating matrix, and various notions of similarity. Depending on the way they considered similarity, the memory-based techniques are grouped into two classes:

- User-based collaborative filtering.
- Item-based collaborative filtering.

The consecutive sections discuss the rudimentary concepts behind the instance- and user-based methods and recall selected implementations and similarity measures.

User-based Collaborative Filtering:

The user-based collaborative filtering works by finding users whose historical ratings are similar to those given by the target user and then using their top-rated products to predict what the target user would like [9]. To mathematically formulate the problem, let us assume there is a list of users $U = \{u_1, u_2, ..., u_m\}$ and a list of items $I = \{i_1, i_2, ..., i_n\}$. Then, the user-item rating matrix consists of a set of ratings $v_{i,j}$ corresponding to the rating for user i on item j. If I_i is the set of items on which user i has rated in the past, then we can define the average rating for user i as follows [9]:

$$\bar{v}_i = \frac{1}{|I_i|} \sum_{j \in I_i} v_{i,j} \tag{5.1}$$

In user-based collaborative filtering, we estimate the rating of item j that has not yet rated by the target user a as follows [9, 13]:

$$p_{a,j} = \bar{v}_a + \frac{\sum_{i=1}^{k} s(a,i)(v_{i,j} - \bar{v}_i)}{\sum_{i=1}^{k} |s(a,i)|} \tag{5.2}$$

where k is the number of most similar users (nearest neighbors) to a. The weights $s(a, i)$ can reflect the degree of similarity between each neighbor i and the target user a.

Item-based Collaborative Filtering:

The item-based collaborative filtering is just an analogous procedure to the collaborative filtering method. The similarity scores can also be used to generate predictions using a weighted average, similar to the procedure used in user-based collaborative filtering. Mathematically, we can predict the rating of item j that has not yet been rated by the target user a as follows [9, 13]:

$$p_{a,j} = \frac{\sum_{i=1}^{k} s(j,i)(v_{a,i})}{\sum_{i=1}^{k} |s(j,i)|} \tag{5.3}$$

where k is the number of most similar items (nearest neighbors) to j that the target user a has rated in the past, and $v_{a,i}$ is the rating given by the target user to every neighbor. Finally, the recommendations are generated by selecting the candidate items with the highest predictions.

Similarity Measures:

The choice of similarity function is considered as a critical design decision while implementing memory-based CF techniques. There are various similarity measures for extracting the set of neighbors, such as cosine similarity adapted from information retrieval and Pearson correlation coefficient [9].

The cosine similarity, or vector similarity, approach is based on linear algebra [13]. In user-based CF, to calculate the similarity between two users, we represent each user as a vector of ratings for the items rated by both users. Then, the similarity is measured by the cosine distance between these two rating vectors. It can be computed efficiently by taking the dot product of these two vectors and dividing it by the product of their L2 (Euclidean) norms (see Eq. 5.4) [13].

$$s(i, j) = \frac{v_i v_j}{||v_i||_2 ||v_j||_2} = \frac{\sum_{k \in I_{i,j}} v_{i,k} v_{j,k}}{\sqrt{\sum_{k \in I_{i,j}} v_{i,k}^2} \sqrt{\sum_{k \in I_{i,j}} v_{j,k}^2}} \tag{5.4}$$

where $I_{i,j}$ represents the items that have been rated by both users i and j.

On the other hand, we can use the Pearson correlation coefficient method to compute the statistical correlation between two user's common ratings, or two item's common ratings, to determine their similarity (see Eq. 5.5) [13].

$$s(i, j) = \frac{\sum_{k \in I_{i,j}} (v_{i,k} - \overline{v}_i)(v_{j,k} - \overline{v}_j)}{\sqrt{\sum_{k \in I_{i,j}} (v_{i,k} - \overline{v}_i)^2} \sqrt{\sum_{k \in I_{i,j}} (v_{j,k} - \overline{v}_j)^2}} \tag{5.5}$$

where $I_{i,j}$ represents the items that have been rated by both users, or items, i and j. Notice that the Eqs. 5.4 and 5.5 can also be utilized to calculate the similarity between two items in item-based CF. In this case, $I_{i,j}$ will represent the number of users who rated both items i and j in the past. Finally, it is worth mentioning that other similarity metrics can also be employed to find the set of neighbors, such as adjusted cosine, constrained Pearson correlation, Euclidean, and mean squared differences [29].

5.2.1.2 Model-Based Techniques

The model-based CF algorithms attempt to discover patterns in the rating history to learn a predictive model in an offline training phase and then use this model online for predicting new ratings [29]. The model building process may be performed by application of various machine learning algorithms, such as matrix factorization (MF) [25, 39, 49], Bayesian classifiers [36, 41], or neural networks and fuzzy systems [47, 57]. Recently, matrix factorization, or latent factors, models have gained popularity due to their ability to address the limitations of memory-based CF algorithms regarding scalability and sparsity issues [33, 56]. Basically, matrix factorization is an unsupervised learning method that is used for dimensionality reduction. One of the most popular techniques applied to solve the matrix factorization problem is Singular Value Decomposition (SVD) [33]. The following section provides a detailed review of the SVD approach as an example of model-based techniques. Also, we show how SVD can be employed in recommendation systems to estimate the ratings.

Singular Value Decomposition (SVD):

SVD technique is used to discover the latent relationships between the users and items in order to factorize the user-item rating matrix into a product of three low-dimensional matrices, which are utilized to calculate the missing ratings in the reduced space [54]. Mathematically, let us assume M is the user-item rating matrix with m users and n items and rank r, SVD is calculated as follows [13, 54]:

$$SVD(M) = U \times S \times V^T \tag{5.6}$$

where U is an orthogonal matrix of size $m \times r$ representing left singular vectors of M where the columns of U are the eigenvectors of MM^T. V is an orthogonal matrix of size $n \times r$ representing right singular vectors of M where the columns of V are the eigenvectors of $M^T M$. S is a diagonal matrix of size $r \times r$, called singular matrix, having r singular values as its diagonal entries which are the positive nonzero eigenvalues of both MM^T and $M^T M$ [13]. Therefore, in order to compute the SVD of matrix M we first compute MM^T and $M^T M$ and then compute the eigenvectors and eigenvalues for MM^T and $M^T M$.

However, it is possible to reduce the $r \times r$ matrix S to have only k largest diagonal values to obtain a matrix S_k, where $k < r$. Accordingly, we need to eliminate $(r - k)$ columns from U and $(r - k)$ rows from V^T since the first r columns of U and V are associated with the r nonzero eigenvalues of MM^T and $M^T M$, respectively. Then, the reconstructed matrix M_k can be produced by multiplying U_k and V_k together using S_k as follows [29, 54]:

$$SVD(M_k) = U_k \times S_k \times V_k^T \tag{5.7}$$

where M_k is the closest rank-k matrix to M, with respect to the Frobenius norm of matrix. Once the reduced matrices (U_k, S_k, V_k) are calculated, we can utilize them to predict the rating for any user u and item i by computing the cosine similarities (dot products) of the cth row of $U_k\sqrt{S_k}$ and the pth column of $\sqrt{S_k}V_k^T$ and adding the user's average ratings as follows [29, 54]:

$$\tilde{r}_{ui} = \bar{r}_u + U_k\sqrt{S_k}(c) \times \sqrt{S_k}V_k^T(p) \tag{5.8}$$

where $\bar{r}_u$ is the average ratings given by user u in the past. c refers to the row index in the resultant matrix $U_k\sqrt{S_k}$ of size $m \times k$, and p refers to the column index of the resultant matrix $\sqrt{S_k}V_k^T$ of size $k \times n$. In the field of recommender systems, SVD can be employed to perform two different tasks [54]:

- Capturing the latent relationships between users and items in order to predict the missing ratings.
- Producing a low-dimensional representation of the original user-item space which allows us to compute neighborhood in the reduced space.

In fact, the low-dimensional representation has several advantages. Firstly, it positively impacts the prediction accuracy by eliminating the small singular values which cause noise in the user-item relationship. Furthermore, it greatly improves the performance of memory-based techniques by making the neighborhood formation process very scalable [29]. Concerning the space storage, we can find that SVD is very efficient since it only stores two matrices of size $m \times k$ and $k \times n$, compared to memory-based techniques which require storing the entire user-item matrix. When it comes to the disadvantages of SVD, in addition to its slowness in the model building process (decomposition step), the conventional SVD is undefined when the original matrix has missing values. This raises difficulties when applying SVD to a sparse rating matrix which is highly prone to overfitting. To deal with this issue, some works relied on imputation, for instance, by replacing the missing values with either the users' or items' average ratings. However, this was not the optimal solution since the data can be distorted by inaccurate ratings. Hence, more recent works suggested only relying on the explicit ratings for model building, while preventing overfitting using the regularization techniques [29, 33].

5.2.2 Content-Based Filtering

(CBF) has become a relevant approach in the development of recommender systems. In contrast to collaborative filtering, content-based filtering does not depend on rating co-occurrences across the users. Instead, it utilizes the characteristics of an item to recommend items similar to those a given user has liked in the past [7]. It is based on the fact that items with similar attributes will be rated similarly. While CBF is very useful, it requires a profound knowledge of each item, in addition to a user profile describing the user's taste. So, the task is to learn the user preferences and then use the most similar items to the ones already evaluated positively by a user to generate a list of recommendations [43]. The recommendation process in content-based filtering is performed in three steps [45]:

- Item profiling.
- User profiling.
- Recommendation generation.

Item profiling, or item representation, is the process of extracting a set of features from a dedicated item. In practice, the item representation can be divided into two main categories:

- Structured.
- Unstructured.

In the structured representation, each item is represented by a set of attributes that have limited and specific values, such as director, actors, and genres in the case of movies. On the other hand, the unstructured representation, such as articles content, the attribute value is unlimited and requires conversion into structured data, such as vectors, on which we can compute metrics, such as Euclidean distance, Pearson's coefficient, or cosine similarity [6].

Generally, content-based filtering techniques are applied to suggest text documents, such as web pages or research articles. In such scenarios, the content of items is represented as text documents, including textual descriptions. Several techniques, such as term frequency and inverse document frequency (TF-IDF), can be used to translate the contents of articles into structured representation so that each item is described by the same set of attributes in the form of a feature vector. TF-IDF uses a set of descriptors or terms to describe the items. These terms can be utilized by other methods, such as vector space model and latent semantic indexing, to represent documents as vectors in a multi-dimensional space [62]. However, TF-IDF can recognize the important terms or phrases of articles. The importance is highly correlated to the popularity of the term in the text, but it decreases its value with the presence of the term in the corpus. If a term is infrequent but its number of occurrences is large in one or a few articles, it probably plays a key role in the article. Basically, TF-IDF is calculated as a combination of the term frequency and inverse document frequency. Term frequency (TF) is the number of times term t appears in document d divided by the document's length. Term Frequency (TF) is defined as follows [46, 62]:

$$TF(t, d) = \frac{N_{t,d}}{N_d} \tag{5.9}$$

where t is a term in the user's query and d is a document in the collection D. $N_{t,d}$ is the number of times that a term t occurs in document d and N_d denotes the number of terms in the document d. Nevertheless, the more popular terms do not give us extra information, and are not useful if they appear in all documents. This is the reason for combining the inverse document frequency (IDF) with the term frequency (TF). IDF reveals how much information the term provides. It is calculated as follows [46, 62]:

$$IDF(t, D) = log\frac{N}{|d \in D : t \in d|} \tag{5.10}$$

where D is a collection of documents and N denotes the number of documents in the collection D. The denominator, $|d \in D : t \in d|$, denotes the number of documents in the collection D that include the term t. Finally, the TF-IDF, or the weight, of term t in document d is calculated as follows [46, 62]:

$$TF\text{-}IDF(t, d, D) = TF(t, d) \times IDF(t, D) \tag{5.11}$$

It can be seen that TF-IDF is proportional to $N_{t,d}$ and inversely proportional to $|d \in D : t \in d|$ [62]. So, the content, or item profile, of text document d for k keywords can be represented in a structured way as a vector as follows:

$$itemProfile(d) = (t_1, t_2, ..., t_k) \tag{5.12}$$

In user profiling (profile learning), the main objective is to generate a profile model for each user based on the previously identified item profiles associated with the items preferred by the associated user [45]. These profiles represent the preference of the user over the set of items. Various methods exist for building user profiles, such as aggregation approaches [1], machine learning-based approaches, or semantic approaches [35].

In the recommendation generation step, we utilize previous profiles to generate the score of the item i for the user u as follows [45]:

$$v(u, i) = score(userProfile(u), itemProfile(i)) \qquad (5.13)$$

The score function represents the relevance of user's profile to item's profile. It can be obtained through similarity functions, such as cosine similarity (see Eq. 5.14) or Jaccard measures [1], as well as semantic similarity approaches [35] and tag-based similarities [38].

$$Similarity = cos(\mathbf{v}_u, \mathbf{v}_i) = \frac{\mathbf{v}_u \mathbf{v}_i}{||\mathbf{v}_u|| ||\mathbf{v}_i||} \qquad (5.14)$$

where $\mathbf{v}_u$ represents the user profile vector, and $\mathbf{v}_i$ represents the feature vector of the candidate item i. Finally, the most relevant N items are recommended to the target user [6, 45].

5.2.3 Challenges and Limitations of CF and CBF

Although CF-based methods are very efficient and give good results, they still suffer some drawbacks. The following are the fundamental challenges to overcome in collaborative filtering recommender systems.

Scalability:

the number of users and items significantly grows over time. Hence it becomes difficult for a recommender system to handle such big data without impacting the latency (waiting time) for each recommendation. This problem is particularly observable in the memory-based collaborative filleting based on, e.g., the k-nearest neighbors. However, making model-based techniques practical and scalable to web-scale recommendation tasks with billions of items and millions of users also remains an unsolved challenge, especially in social media, the Internet of Things (IoT), or various e-commerce applications like fast-moving consumer goods industries [20, 37]. Among the popular methods often used to address this issue, we may mention soft clustering and bi-clustering, or dimensionality reduction techniques like SVD and latent factor models/matrix factorization. In [63], the authors present a distributed framework that can scale different association-rule-based recommendation methods, which handle the ever-increasing number of online customers by a distributed framework with two load-balanced strategies for sparse and dense data, respectively. For scaling up the semantics-driven RS, the pre-computation of the cosine similarities and gradient learning of the model may be applied [8]. In [64], the authors develop a data-efficient Graph Convolutional Network (GCN) algorithm and efficient MapReduce model inference algorithm to solve the related issue, which is especially popular considering the low-cost cloud resources [21]. In the context of IoT, the development of edge computing plays an increasingly important role in the optimization or recommendation latency [26].

Sparsity:

since CF-based methods rely on modeling the user-item interactions, the recommendation quality of these methods may get impacted by an insufficient number of items rated by each user. Sometimes, the similarity between each pair of users cannot be computed due to the lack of ratings given to common items. Thus, making the user-based collaborative filtering algorithm useless in such scenarios [28, 56, 59].

Cold-start problem:

it occurs whenever a recommender system tries to generate recommendations for either a new user who signed up recently to the system without having any rating records available yet or when a new item is added to the system without any rating given to that item so far. In fact, most state-of-the-art recommendation algorithms generate unreliable recommendations for such cases since they cannot learn the preference embedding of these new users/items [34, 59].

The adoption of the content-based filtering approach has several important advantages when compared to the collaborative approach. These advantages are summarized as follows:

User Independence/Privacy:

In content-based filtering approach, there is no need to use the ratings of other users (neighbors) to generate the recommendations like in collaborative filtering methods. Instead, CBF provides user independence by using exclusive ratings provided by the active user to build their own profile. This preserves the users' privacy by avoiding utilizing their ratings to generate recommendations for other users [19, 42, 58].

Transparency/Explainable Recommendations:

Content-based filtering can provide quite explainable recommendations by explicitly listing content features that caused an item to be recommended which will potentially increase the users' trust and confidence in the output of the recommender system.

In contrast, the only explanations provided by collaborative filtering methods to justify some recommended items is that other users with similar tastes like these items [19, 42, 58].

Item cold-start problem:

Content-based filtering methods are capable to recommend new items which have been added recently to the system and most likely do not have, or have very few, ratings in the past. Conversely, the recommendations generated by collaborative filtering techniques are negatively impacted in this scenario since they solely rely on the historical ratings [34, 42, 58].

Dynamic evolving:

Content-based techniques are capable to adjust the recommendations quickly in response to the dynamics of evolving user preferences [42].

Nonetheless, content-based filtering poses a number of drawbacks and limitations though:

Limited content analysis:

Extract plenty and very well descriptive attributes is necessary in order to distinguish between the items. However, it is difficult to generate sufficient attributes to define distinguishing aspects of items in certain areas [19, 42, 58].

Over-specialization (serendipity problem):

Content-based filtering recommender systems suggest items that have high scores in matching the user profile with no possibility to recommend new different and interesting items for the target user. In fact, content-based filtering methods suffer from the serendipity problem, when the recommender system does not have the tendency to produce recommendations with a limited degree of novelty [19, 42].

User cold-start problem:

In content-based recommender systems, it is necessary to learn the user preferences in order to provide reliable recommendations. Therefore, CBF suffers from the user cold-start problem when new users who signed up recently do not have, or have very few, ratings, and hence, the quality of recommendation will be impacted by an insufficient number of rated items [19, 42].

The above-mentioned challenges and limitations of CF and CBF prompted researchers to develop hybrid methods combining both approaches.

5.2.4 Hybrid Filtering

The model is a combination of more than one filtering approach. It is introduced to overcome some common problems that are associated with filtering approaches, such as the cold-start problem, over-specialization problem, and sparsity problem [32]. For example, the sparsity problem is addressed in the hybrid recommendation model by integrating the information of the content-based filtering and collaborative filtering models. Another motivation behind the implementation of hybrid filtering is to improve the accuracy and efficiency of the recommendation process. However, the outcomes of any hybrid model strongly depend on the used algorithms and the method of hybridization, for example, how the outcomes of an algorithm relate to the other ones [42]. Basically, hybrid methods can be implemented in various ways as follows [58]:

- Implement collaborative and content-based methods individually, then aggregate their predictions to yield better recommendations.
- Integrate some content-based characteristics into a collaborative method. This greatly helps to address the cold start problem in collaborative filtering and the over-specialization problem of content-based filtering.
- Integrate some collaborative characteristics into a content-based method.
- Construct a single unified recommendation model that integrates both content-based and collaborative characteristics to improve the effectiveness of the recommendation process.

Finally, a comprehensive list of the various hybridization techniques has been summarized in Table 5.1.

5.2.5 Evaluation of Recommendation Systems

There are three main ways to assess RS quality: user studies, online, and offline valuation. User studies collect feedback from a limited number of real system users. Online experiments evaluate the real-time use of RS systems, whereas offline experiments rely on historical data to replicate user activities and measure RS quality [31]. This section reviews the most popular offline evaluation methods, which allow us to assess and optimize system quality before deploying it to production environments.

In literature, we may find a vast number of evaluation strategies to assess recommender systems' quality that incorporate various protocols, data splitting techniques, quality metrics, and significance testing of the results achieved. The application of the chosen approach depends on the investigated recommendation problem and the nature of the data. Most evaluation strategies adhere to those well-known from machine learning and assume splitting the data into training and testing sets. The first simulates the already collected historical data that is used to train and tune the RS. The latter is used to assess its performance [12]. There are many methods to split data

Table 5.1 Seven types and concepts of hybridization methods [3, 10, 32]

Hybrid method	Description
Weighted	In this method, the weight of each recommender system is gradually adjusted according to the degree to which the user's evaluation of an item coincides with the evaluation predicted by the recommender system. Finally, the scores of several recommendation techniques are statistically aggregated to produce a single recommendation
Switching	The system switches between different recommendation techniques and uses the one expected to have the best result in a particular/current context
Mixed	Recommendations from more than one technique are presented at the same time
Feature combination	Features from different recommendation data sources are combined and used as inputs to a single recommender system. For example, injecting features of the collaborative filtering model (sub-RS) into the content-based model (primary-RS)
Feature augmentation	The output of one recommender system is used to create input features (augmented dataset) for the next one. For example, a sub-RS is employed to generate a rating of the user/item profile, which is further used in the primary recommender system to produce the final predicted result
Cascade	Similar to boosting technique, the cascade method consecutively refines the recommendations of the previous system, and the overall results are combined into a single output
Meta-level	The model learned by one recommender system is used as input to another one. In this method, the sub-RS replaces the original dataset with a learned model which is further used as the input to the primary recommendation model

into training and testing sets like random sampling or K-fold cross-validation, and slightly different approaches like data lagging in case of time-aware recommendation problems [5, 48]. There may also be certain modifications of the data-splitting procedure when evaluating a RS against particular challenges like the cold-start problem related to the emergence of new items or users [15].

One of the key decisions is the appropriate selection of the quality metric that implies which quality aspect of the recommendation system we will optimize [55]. Below we present an overview of the popular quality measures.

Root Mean Squared Error (RMSE):

It is one of the most popular metrics used in evaluating the accuracy of predicted ratings. It imposes a penalty over the larger errors [6]. The between the predicted and actual ratings is given by:

$$RMSE = \sqrt{\frac{1}{N} \sum_{i=1}^{N} (r_i - \hat{r}_j)^2} \qquad (5.15)$$

where N is the number of ratings in the test set. r_i is the true rating value and $\hat{r}_j$ is the predicted rating.

Mean Absolute Error ():

This metric is also used to evaluate the accuracy of the prediction made by recommendation algorithms [6]. It measures the average magnitude of the errors in a set of predictions, without considering their direction. it can be calculated by the following equation:

$$MAE = \frac{1}{N} \sum_{i=1}^{N} |r_i - \hat{r}_i| \tag{5.16}$$

The lower the MAE or RMSE, the better system predicts the ratings.

Precision:

It quantifies the number of correct positive recommendations made. The bigger precision value indicates the more accurate recommendation that the recommender system made [6, 22]. Its the equation is as follows:

$$Precision = \frac{TP}{TP + FP} = \frac{\text{Relevant recommendations}}{\text{Total recommendations}} \tag{5.17}$$

Recall:

It quantifies the number of correct positive recommendations made out of all positive predictions that could have been made. The bigger recall value means that the recommendation system has the ability to rank the most relevant papers at the top of the result list [6, 22]. Its equation is as follows:

$$Recall = \frac{TP}{TP + FN} = \frac{\text{Relevant recommendations}}{\text{Total relevant recommendations}} \tag{5.18}$$

F-Measure:

It provides a way of combining the precision and recall into a single measure that captures both properties [6, 22]. F-measure is commonly used when the recommender system feedback is binary information: recommend or not recommend. it is defined as the harmonic mean of the model's precision and recall as follows:

$$F_1 = 2 \cdot \frac{precision \cdot recall}{precision + recall} \tag{5.19}$$

Accuracy:

It is used to measure all the correctly identified cases. This measure is mostly used when all the classes are equally important as follows:

$$Accuracy = \frac{TP + TN}{TP + FP + TN + FN} \qquad (5.20)$$

(Normalized Discounted Cumulative Gain):

It is used to evaluate the quality of a given sorted list of recommendations. It depends on the Discounted Cumulative Gain () which assumes that highly relevant items that appear at the end of the recommendations list should be penalized [45]. DCG is formalized as follows:

$$DCG_u = \sum_{k=1}^{N} \frac{r_{u,recom_k}}{\log_2 k + 1} \qquad (5.21)$$

where N is the number of recommended items to user u, and $recom_k$ is the item recommended to user u in k position. Notice that the graded relevance value is reduced logarithmically in proportion to the position of the result. To calculate NDCG, we normalize the DCG value by dividing it by the maximum DCG, $DCG_{perfect}$, where the most relevant items appear first on the list [45]. The NDCG value for each user is calculated as follows:

$$NDCG = \frac{DCG}{DCG_{perfect}} \qquad (5.22)$$

User Coverage (UCOV):

It measures the percentage of users who received relevant recommendations, Its equation is as follows:

$$UCOV = \frac{\mathcal{U}^{\cdot}}{\mathcal{U}} \qquad (5.23)$$

where $\mathcal{U}^{\cdot}$ is the number of users who get relevant recommendations, and $\mathcal{U}$ is the number of all the users in the system.

All evaluation metrics we discussed so far are considered as accuracy-based measures since they focus on measuring the quality and effectiveness of recommendations. Although accuracy-based measures are very important, there are traits of user satisfaction they are unable to capture. Recently, many researchers and practitioners have noticed that accuracy is not the only criteria of interest, hence they started focusing on beyond accuracy measures, including diversity, novelty, serendipity (surprise), coverage, and many other metrics [66]. Although these metrics are not as common yet in RS evaluation, they attracted a lot of attention recently due to their importance in many recommendation applications. Next, we introduce and discuss some beyond accuracy measures.

Diversity:

It is generally defined as the opposite of similarity. This measure shows how dissimilar are the recommendations provided for the user. This similarity is often determined using the item's content, or it can be determined using how similarly items are rated. The most frequently considered diversity metric is Intra-List Similarity (). ILS of a set of recommended items is defined as the pairwise distance of the items in the set [66]. It is formalized as follows:

$$ILS(R) = \frac{1}{|R|(|R| - 1)} \sum_{i \in R} \sum_{j \in R, i \neq j} d(i, j) \tag{5.24}$$

where $d(i, j)$ represents the distance measure, such as Hamming distance, Euclidean distance, Cosine similarity, Pearson correlation coefficient, and Jaccard similarity coefficient.

Novelty:

It measures how novel are the items based on their popularity. The popularity is modeled as the probability that a random user would know the item, hence an item is novel if few people are aware it exists. Thus, item novelty could be evaluated by the Inverse User Frequency () as follows [60]:

$$IUF = NOV(i) = \log_2 \frac{|\mathcal{U}|}{|\mathcal{U}_i|} \tag{5.25}$$

where $\mathcal{U}_i$ is the users who have rated item i and $\mathcal{U}$ represents the total number of users. The main idea of IUF is to penalize the predicted rating of popular items and promote unpopular ones. From a business point of view, this metric is used to increase the sales of long-tail (unpopular) products [16]. Thus, the novelty of a recommendation can be assessed as the average IUF of the recommended items, which is called Mean Inverse User Frequency (MIUF) [65]:

$$MIUF(R) = -\frac{1}{|R|} \sum_{i \in R} \log_2 \frac{|\mathcal{U}_i|}{|\mathcal{U}|} \tag{5.26}$$

where R is the set of recommended items for one user. $\mathcal{U}_i$ is the set of users who have interacted with item i, and $\mathcal{U}$ represent all users in the system.

5.3 Factorization Machine and Association Rules (FMAR) Recommender System

In this section, we formally provide the statement of the problem that we aim to tackle. After that, we briefly summarize the academic knowledge of the techniques employed in . Then, we introduce the details of a novel recommender system that is based on Factorization Machine and Association Rules (FMAR). The proposed model has two versions based on the algorithm used to generate the association rules:

- Factorization machine apriori based model.
- Factorization machine FP-growth based model.

5.3.1 Problem Definition

In many recommender systems, the elapsed time required to generate the recommendations is very crucial. Moreover, in some systems, any delay in generating the recommendations can be considered as a failure in the recommendation engine. The main problem we address in this paper is to minimize the prediction latency of the recommender system by incorporating the association rules in the process of building the recommender system. The main idea is to use the association rules to decrease the number of items that we need to approximate their ratings, hence, decreasing the time that the recommender system requires to generate the recommendations. Also, our goal is to make sure that the accuracy of the final recommendations is not impacted after filtering the items using the association rules.

5.3.1.1 Factorization Machines

In linear models, the effect of one feature depends on its value. While in polynomial models, the effect of one feature depends on the value of the other features. Reference [50] can be seen as an extension of a linear model which efficiently incorporates information about features interactions, or it can be considered equivalent to polynomial regression models where the interactions among features are taken into account by replacing the model parameters with factorized interaction parameters [18].

However, polynomial regression is prone to overfitting due to a large number of parameters in the model. Needless to say, it is computationally inefficient to compute weights for each interaction since the number of pairwise interactions scales quadratically with the number of features. On the other hand, factorization machines elegantly handled previous issues by finding a one-dimensional vector of size k for each feature. Then, the weight values of any combination of two features can be represented by the inner product of the corresponding features vectors. Therefore, factorization machines manage to factorize the interactions weight matrix $W \in R^{n \times n}$, which is used in polynomial regression, as a product $V V^T$, where $V \in R^{n \times k}$. So, instead of modeling all interactions between pairs of features by independent parameters like in polynomial regression (Eq. (5.27)):

$$\widehat{y}(\mathbf{x}) = w_0 + \sum_{i=1}^{n} w_i x_i + \sum_{i=1}^{n} \sum_{j=i+1}^{n} w_{ij} x_i x_j \qquad (5.27)$$

where:

$\mathbf{x} = (x_1, \cdots, x_n) \in \mathbb{R}^n$: is a n dimension input vector

$w_0 \in \mathbb{R}$: is the global bias.

$w_i \in \mathbb{R}^n$: models the strength of the i-th variable.

$w_{ij} \in \mathbb{R}^{n \times n}$: models the interaction between the ith and j-th variable.

We can achieve that using factorized interaction parameters, also known as latent vectors $\mathbf{v}_1, \cdots, \mathbf{v}_n \in \mathbb{R}^k$, in factorization machines (Eq. (5.28)).

$$\widehat{y}(\mathbf{x}) = w_0 + \sum_{i=1}^{n} w_i x_i + \sum_{i=1}^{n} \sum_{j=i+1}^{n} \langle \mathbf{v}_i^T \mathbf{v}_j \rangle x_i x_j \qquad (5.28)$$

where k is a predefined parameter and $\mathbf{v}_i^T \mathbf{v}_j$ is a scalar resulting form the dot product of two vectors of size k, namely: $\mathbf{v}_i$ and $\mathbf{v}_j$. Intuitively, k-dimension latent vector $\mathbf{v}_i$ describes some hidden characters of the feature x_i for $i = 1, \cdots, n$ and $\mathbf{v}_i^T \mathbf{v}_j$ models the interaction between the i-th and j-th variable by factorizing it.

In case of the recommendation task, each vector $\mathbf{x}$ represents a pair of user, item and the context the user rates the item, while $\widehat{y}(\mathbf{x})$ is the prediction of the rating for this pair. The learning problem is to search for the parameters $w_0, w_1, \cdots w_n$ and matrix $V = [\mathbf{v}_1, \cdots, \mathbf{v}_n] \in \mathbb{R}^{n \times k}$ that minimizes the different between $\widehat{y}(\mathbf{x})$ and the real rating $r(\mathbf{x})$.

The most important feature of the factorization machine is based on the fact that even the Eq. (5.28) is spanning on a pairwise space [40] and it requires $O(kn^2)$ computational complexity, it can be effectively calculated in $O(nk)$ time, i.e. linear time because k is a constant selected by the user. In fact:

$$y(\mathbf{x}) = \sum_{i=1}^{n} \sum_{j=i+1}^{n} \langle \mathbf{v}_i^T \mathbf{v}_j \rangle x_i x_j \qquad (A)$$

$$= \sum_{1 \leq i < j \leq n} \langle \mathbf{v}_i^T \mathbf{v}_j \rangle x_i x_j = \frac{1}{2} \sum_{i=1}^{n} \sum_{j=1 \text{ and } j \neq i}^{n} \langle \mathbf{v}_i^T \mathbf{v}_j \rangle x_i x_j$$

$$= \frac{1}{2} \sum_{i=1}^{n} \sum_{j=1}^{n} \langle \mathbf{v}_i^T \mathbf{v}_j \rangle x_i x_j - \frac{1}{2} \sum_{i=1}^{n} \langle \mathbf{v}_i^T \mathbf{v}_i \rangle x_i x_i \qquad (B)$$

This transformation is simply an index maneuvering, as in Eq. (A) we have a sum of terms $\langle \mathbf{v}_i^T \mathbf{v}_j \rangle x_i x_j$ for all $1 \leq i < j \leq n$. As $\langle \mathbf{v}_i^T \mathbf{v}_j \rangle = \langle \mathbf{v}_j^T \mathbf{v}_i \rangle$, this sum can be rewritten in by Eq. (B). Now, lets assume that $\mathbf{v}_i = [v_{i,1}, \cdots, v_{i,k}]^T \in \mathbb{R}^k$ for $i = 1, \cdots, n$, then $\langle \mathbf{v}_i^T \mathbf{v}_i \rangle = \sum_{f=1}^{k} v_{i,f} v_{j,f}$. Thus

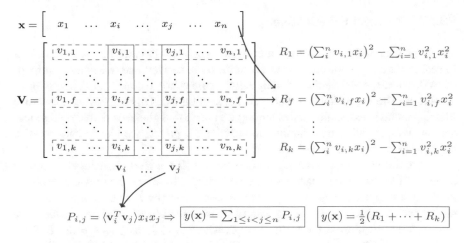

Fig. 5.1 The illustration of the transformation from the quadratic computation (Eq. (A)) to the linear (computation Eq. (E))

$$
\begin{aligned}
y(\mathbf{x}) &= \frac{1}{2}\left(\sum_{i=1}^{n}\sum_{j=1}^{n}\sum_{f=1}^{k} v_{i,f}v_{j,f}x_i x_j\right) - \frac{1}{2}\left(\sum_{i=1}^{n}\sum_{f=1}^{k} v_{i,f}v_{i,f}x_i x_i\right) \\
&= \frac{1}{2}\left(\sum_{i=1}^{n}\sum_{j=1}^{n}\sum_{f=1}^{k} v_{i,f}v_{j,f}x_i x_j - \sum_{i=1}^{n}\sum_{f=1}^{k} v_{i,f}v_{i,f}x_i x_i\right) \quad\quad\quad (C) \\
&= \frac{1}{2}\sum_{f=1}^{k}\left(\sum_{i=1}^{n}\sum_{j=1}^{n} v_{i,f}v_{j,f}x_i x_j - \sum_{i=1}^{n} v_{i,f}v_{i,f}x_i x_i\right) \quad\quad\quad\quad (D) \\
&= \frac{1}{2}\sum_{f=1}^{k}\left(\left(\sum_{i=1}^{n} v_{i,f}x_i\right)\left(\sum_{j=1}^{n} v_{j,f}x_j\right) - \sum_{i=1}^{n} v_{i,f}^2 x_i^2\right) \\
&= \frac{1}{2}\sum_{f=1}^{k}\left(\left(\sum_{i}^{n} v_{i,f}x_i\right)^2 - \sum_{i=1}^{n} v_{i,f}^2 x_i^2\right) \quad\quad\quad\quad\quad (E)
\end{aligned}
$$

The transformation from (C) to (D) is changing the order of summation. The whole idea is illustrated in Fig. 5.1.

This advantage is very useful in recommendation systems since the datasets are mostly sparse, and this will adversely affect the ability to learn the feature interactions matrix as it depends on the feature interactions being explicitly recorded in the available dataset.

5.3.1.2 Frequent Pattern Mining

The basic idea of frequent pattern mining, also known as association rule mining, is to search for all relationships between elements in a given massive dataset. It helps us to discover the associations among items using every distinct transaction in large databases. The key difference between association rules mining and collaborative filtering is that in association rules mining we aim to find global or shared preferences across all users rather than finding an individual's preference like in collaborative filtering-based techniques [14, 23, 27].

At a basic level, association rule mining analyzes the dataset searching for frequent patterns (itemsets) using machine learning models. To define the previous problem mathematically, let $I = \{i_1, i_2, ..., i_m\}$ be an itemset and let D be a set of transactions where each transaction T is a nonempty itemset such that $T \subseteq I$. An association rule is an implication of the form $A \Rightarrow B$, where $A \subset I$, $B \subset I$, $A \neq \varnothing$, $B \neq \varnothing$, $A \cap B = \varnothing$. In the rule $A \Rightarrow B$, A is called the antecedent and B is called the consequent. Various metrics are used to identify the most important itemset and calculate throe strength, such as support, confidence, and lift. Support metric [14] is the measure that gives an idea of how frequent an itemset is in all transactions. In other words, the support metric represents the number of transactions that contain the itemset. The Eq. 5.29 shows how we calculate the support for an association rule.

$$support(A \Rightarrow B) = P(A \cup B) \tag{5.29}$$

On the other hand, the confidence [14] indicates how often the rule is true. It defines the percentage of transactions containing the antecedent A that also contain the consequent B. It can be taken as the conditional probability as shown in Eq. 5.30.

$$confidence(A \Rightarrow B) = P(B|A) = \frac{support(A \cup B)}{support(A)} \tag{5.30}$$

Finally, the lift is a correlation measure used to discover and exclude the weak rules that have high confidence. The Eq. 5.31 shows that the lift measure is calculated by dividing the confidence by the unconditional probability of the consequent [14, 23].

$$lift(A \Rightarrow B) = \frac{P(A \cup B)}{P(A)P(B)} = \frac{support(A \cup B)}{support(A)support(B)} \tag{5.31}$$

If the lift value is equal to 1, then A and B are independent and there is no correlation between them. If the lift value is greater than 1, then A and B are positively correlated. If the lift value is less than 1, then A and B are negatively correlated.

Various algorithms exist for mining frequent itemsets, such as Apriori [4, 27], AprioriTID [4, 27], Apriori Hybrid [4, 27], and FP-growth (Frequent pattern) [24, 27]. In this paper, we employ FP-growth algorithm to generate frequent itemsets. What makes FP-growth better than other algorithms is the fact that FP-growth algorithm relies on FP-tree (frequent pattern tree) data structure to store all data concisely and compactly which greatly helps to avoid the candidate generation step. Moreover, once the FP-tree is constructed, we can directly use a recursive divide-and-conquer approach to efficiently mine the frequent itemsets without any need to scan the database over and over again like in other algorithms [24].

5.3.2   Factorization Machine Apriori Based Model

In this section, we introduce the reader to the first version of FMAR which proposes a hybrid model that utilizes factorization machines [50] and apriori [4, 27] algorithms to speed up the process of generating the recommendations. Firstly, we use to create a set of association rules based on the rating history of users. Secondly, we use these rules to create users' profile which recommends a set of items for every user. Then, when we need to generate recommendations for a user, we find all products that are not rated before by this user, and instead of generating predictions for all of them, we filter them using the items in the users' profile. Finally, we pass the short-listed set of items to a recommender system to estimate their ratings using factorization machines model.

In the context of association rules, it is worth noting that while generating the rules, all unique transactions from all users are studied as one group. On the other hand, while calculating the similarity matrix in collaborative filtering algorithms, we need to iterate over all users and every time we need to identify the similarity matrix using transactions corresponding to a specific user. However, what we need to do to improve the recommendation speed is to generate predictions for parts of items instead of doing that for all of them. Next, we introduce the algorithms that we use to generate the association rules and users' profile.

After generating the users' profile, we can use it to improve the recommendation speed for any user by generating predictions for a subset of not-rated items instead of doing that for all of them. The filtering is simply done using the recommended items which are extracted for every user using the association rules as shown in Algorithm 2. Moreover, the filtering criteria can be enhanced by using the recommended items of the closest n-neighbors of the target user. The similarity between the users can be calculated using pearson correlation or cosine similarity measures.

5.3.3   Factorization Machine FP-Growth Based Model

In this section, we introduce the second version of FMAR where [24, 27] algorithm has been employed to generate the association rules. In general, FP-growth algorithm is considered as an improved version of apriori method which has two major shortcomings:

Algorithm 1 Association Rules Generation Using Apriori Algorithm

---
1: Extract favorable reviews   ▷ ratings > 3
2: Find frequent item-sets $\mathcal{F}$ ▷ support > min_support
3: Extract all possible association rules $\mathcal{R}$
4: **for** $r \in \mathcal{R}$ **do**
5:     Compute confidence and lift
6:     **if** (confidence < min_confidence) or (lift < min_lift) **then**
7:         Filter out this rule from $\mathcal{R}$
8:     **end if**
9: **end for**
10: Create users' profile using rules in $\mathcal{R}$

Algorithm 2 Users' Profile Generation

1: **for** each user **do**
2:     Find high rated items based on rating history
3:     Find the rules that their antecedents are subset of high rated items
4:     Recommend all consequences of these rules
5: **end for**

- Candidate generation of itemsets which could be extremely large.
- Computing support for all candidate itemsets which is computationally inefficient since it requires scanning the database many times.

However, what makes FP-growth algorithm different from apriori algorithm is the fact that in FP-growth no candidate generation is required. This is achieved by using FP-tree (frequent pattern tree) data structure which stores all data in a concise and compact way. Moreover, once the FP-tree is constructed, we can directly use a recursive divide-and-conquer approach to efficiently mine the frequent itemsets without any need to scan the database over and over again. Next, we introduce the steps followed to mine the frequent itemsets using FP-growth algorithm. We will divide the algorithm into two stages:

- FP-tree construction.
- Mine frequent itemsets.

After finding the frequent itemsets, we generate the association rules and users' profiles in the same way as in FM Apriori-based model. Finally, in order to generate predictions, we employ a factorization machines model, which is created using the publicly available software tool libFM [51]. This library provides an implementation for factorization machines in addition to proposing three learning algorithms: stochastic gradient descent (SGD) [50], alternating least-squares (ALS) [52], and Markov chain Monte Carlo (MCMC) inference [17].

Algorithm 3 FP-Tree Construction

1: Find the frequency of 1-itemset
2: Create the root of the tree (represented by null)
3: **for** each transaction **do**
4:    Remove the items below min_support threshold
5:    Sort the items in frequency support descending order
6:    **for** each item **do**   ▷ starting from highest frequency
7:       **if** item not exists in the branch **then**
8:          Create new node with count 1
9:       **else**
10:          Share the same node and increment the count
11:       **end if**
12:    **end for**
13: **end for**

Algorithm 4 Mining Frequent Itemsets

1: Sort 1-itemset in frequency support ascending order
2: Remove the items below min_support threshold
3: **for** each 1-itemset **do**   ▷ starting from lowest frequency
4:    Find conditional pattern base by traversing the paths in FP-tree
5:    Construct conditional FP-tree from conditional pattern base
6:    Generate frequent itemsets from conditional FP-ree
7: **end for**

5.4   Experimental Evaluation of FMAR

In this section, we conduct comprehensive experiments to evaluate the performance of the FMAR recommender system. In our experiments, we used MovieLens 100 K dataset[1] which was collected by the GroupLens research project at the University of Minnesota. MovieLens 100 K is a stable benchmark dataset that consists of 1682 movies and 943 users who provide 100, 000 ratings on a scale of 1 to 5. It is important to note that, in this paper, we are not concerned about users' demographics and contextual information since the association rules are generated based only on rating history.

In order to provide a fair comparison, we refer to several metrics and methods to evaluate FMAR and FM recommender systems, such as Mean Absolute Error (MAE), Normalized Discounted Cumulative Gain (NDCG), and Wilcoxon Rank-Sum test.

To construct the test set that allows us to verify both the quality and performance/scalability of the RS, we sampled 50 users with a significant amount of ratings in the data. For those users, 70% of their ratings constituted the testing set. The training set was constructed using the rest of the records in the data (not used in testing sets). The training set was used to generate the association rules and build the factorization machines model. Furthermore, we created three versions of the testing

---

[1] https://grouplens.org/datasets/movielens/.

set. The first one, called *original*, contains all the testing data. While the second and third, called *short-listed*, were created by filtering out possibly many of the records from the original test set using the association rules generated by the two developed methods, i.e., Apriori-based and FP-growth-based FM models. Finally, we pass those sets to the factorization machines model to generate predictions and evaluate both versions of FMAR, by comparing them with the standard FM model operating on the complete data.

In the first experiment, we calculate the mean absolute error (MAE) generated in both recommendation engines. The main goal of this approach is to show that the quality of recommendations is not significantly impacted after filtering the items in the testing set using the association rules. Fig 5.2(left) compares the mean absolute error of the predictions made using FM model with FM Apriori model and FM FP-growth model for 50 users. We use a box plot, which is a standardized way of displaying the distribution of data, to plot how the values of mean absolute error are spread out for 50 users. This graph encodes five characteristics of the distribution of data which show their position and length. These characteristics are minimum, first quartile (Q1), median, third quartile (Q3), and maximum. The results of this experiment show that MAE of FMAR in both versions, FM Apriori and FM FP-growth, is very close to MAE of FM recommender system. However, the average value of MAE for 50 users is 0.71 using FM model, 0.80 using FMAR (Apriori model), and 0.78 using FMAR (FP Growth model).

In the second experiment, we evaluate FMAR by comparing its recommendation with FM using Normalized Discounted Cumulative Gain (NDCG) which is a measure of ranking quality that is often used to measure the effectiveness of recommendation systems or web search engines. NDCG is based on the assumption that highly relevant recommendations are more useful when appearing earlier in the recommendations list. So, the main idea of NDCG is to penalize highly relevant recommendations that appear lower in the recommendation list by reducing the graded relevance value logarithmically proportional to the position of the result. Figure 5.2 (right) compares the accuracy of FM model with FM Apriori model and FM FP-growth model for 50 users and shows the distribution of the results. It is worth noting that in this test we calculate NDCG using the highest 10 scores in the ranking which are generated by FM or FMAR. The results show that both versions of FMAR model have always higher NDCG values than FM model which means that true labels are ranked higher by predicted values in FMAR model than in FM model for the top 10 scores in the ranking.

In the third evaluation method, we run Wilcoxon Rank-Sum test on the results of previous experiments. Firstly, we apply Wilcoxon Rank-Sum test to the results of the first experiment in order to prove that the difference in MAE between FM and FMAR is not significant, and hence, it can be discarded. So, we pass two sets of samples of MAE for FM and FMAR. The Table 5.2 shows the p-value for comparing FMAR using Apriori model and FP Growth model with FM model. In both cases, we got p-value > 0.05 which means that the null hypothesis is accepted at the 5% significance level, and hence, the difference between the two sets of measurements is not significant. On the other hand, we apply Wilcoxon Rank-Sum test to the results of

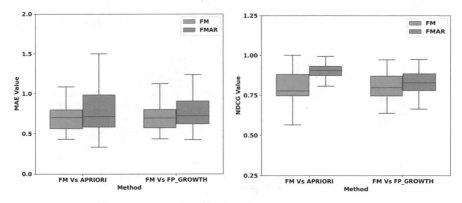

Fig. 5.2 MAE Comparison (left) an NDCG Comparison (right)

Table 5.2 Wilcoxon Rank-Sum test

Model	p-value
MAE (FM Vs FMAR-Apriori model)	0.29
MAE (FM Vs FMAR-FP Growth model)	0.13
NDCG (FM Vs FMAR-Apriori model)	1.74e-08
NDCG (FM Vs FMAR-FP Growth model)	0.04

the second experiment to check if the difference in NDCG between FM and FMAR is significant. However, the Table 5.2 shows that p-value < 0.05 for comparing both models of FMAR with FM model. This means the null hypothesis is rejected at the 5% significance level (accept the alternative hypothesis), and hence, the difference is significant. Since FMAR model has higher NDCG than FM model, we can conclude that FMAR outperforms FM for the highest top 10 predictions.

In the next step, we verified the impact of limiting the amount of data with association rules on the diversity of the same top 10 recommendations that were evaluated with NDCG. For that purpose, we referred to a genre (category) of the movies to calculate the distance between them. We assessed the diversity of the resulting set of 10 recommended movies with the Intra-List Similarity (Eq. 5.24). For example, if two films belong to the "Action" and "Comedy" categories, then ILS = 2. This experiment revealed that the macro averaged ILS of the FP-Growth version of FMAR is equal to 0.32 and is slightly better when compared to the FM, for which ILS equals 0.33. However, the difference was not statistically significant. However, the ILS score for Apriori-based FMAR was significantly higher (i.e., ILS = 0.44), i.e., a p-value of the Wilcoxon rank-sum test equals $7.72e$-11. For the investigated data, the Apriori-based association rule filtering resulted in the more similar yet more optimized recommendations (cf. the narrow box plot in Fig. 5.2 (right)), whereas the FP-Growth version preserves the original diversity.

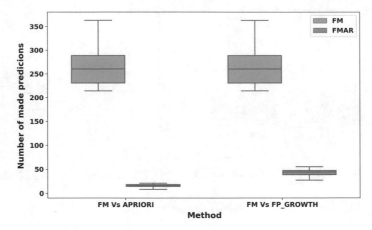

Fig. 5.3 Comparison of the speed of methods (estimated by the number of predictions made by the factorization machine model, i.e., the lower the better)

In the last experiment, we compare FM and FMAR in terms of the speed of their operation, measured as the number of predictions performed by the factorization machines model. The main idea is to estimate the time necessary to prepare recommendations for every tested user for both evaluated approaches. Figure 5.3 shows the distribution of the results of this experiment for the selected 50 test users. Observably, the number of items that we need to predict with FMAR is significantly lower due to using the association rules for filtering. The results show that the FMAR model can generate predictions for any user at least four times faster than the FM model. However, it is noteworthy that generating the rules is part of the training procedure. Therefore, it is a one-time effort, and there is no need to regenerate or update the association rules frequently in FMAR. Therefore, the computational cost of the training procedure for every method, including extracting the association rules, is not considered in our comparisons.

On the other hand, it is worth mentioning that the association rules in our experiments are generated using the entire dataset which means that these rules try to find global or shared preferences across all users. However, another way to generate the association rules is to split the dataset based on users' demographics, such as gender, or items' characteristics, such as genre, or even contextual information, such as weather and season. Thus, if we are producing recommendations for a female user in winter season, we can use dedicated rules which are extracted from historical ratings given by females in winter season. Following this strategy, we can generate multiple sets of recommendation rules which can be used later during prediction time to filter the items. Obviously, the rules generated after splitting the dataset will be smaller. So, the prediction latency can be minimized by selecting a smaller set of rules. In fact, this feature is very useful when we want to make a trade-off between the speed and quality of recommendations. Lastly, it is important to note that several experiments are conducted in order to select the appropriate values of hyper-parameters used in previous algorithms. For instance, min_support = 250, min_confidence = 0.65, number of epochs in FM = 100, number of factors in FM = 8. However, multiple factors are taken into consideration while selecting those values, including accuracy, number of generated rules, and memory consumption.

5.5 Summary

The chapter reviews various approaches to building recommendation systems and presents an overview of the most prominent techniques related to collaborative and content filtering as well as hybrid methods. The chapter discusses typical applications of RSs and summarises evaluation methods adequate to particular recommendation tasks. Furthermore, this chapter introduces FMAR, a novel recommender system, which methodically incorporates the association rules in generating the recommendations using the factorization machines model. The presented study evaluates two approaches to creating association rules based on the users' rating history, namely: the apriori and frequent pattern growth algorithms. These rules are used to decrease the number of items passed to the model to estimate the ratings and hence, reduce the recommender system's prediction latency. We presented extensive experimental research examining the speed of generating recommendations, the quality of estimating user ratings, the quality of selecting the most relevant recommendations, and their diversity. The results confirmed that FMAR has significantly improved the efficiency of recommender systems with no negative impact on the quality and diversity.

In the future work, we plan to incorporate more information in producing the association rules, including users' and items' content-based characteristics and contextual information. Another vital aspect is evaluating our proposed model against more state-of-the-art approaches and datasets. Furthermore, we plan to consider the drifts and shifts in data, like changes in users' behavior and preferences. Another direction of future work is to utilize the generated association rules to solve the cold-start problem. Finally, we plan to introduce more distributed stream processing into our algorithms.

Acknowledgements Research supported by Polish National Science Centre (NCN) grant no. 2018/31/N/ST6/00610.

References

1. Adomavicius, G., Tuzhilin, A.: Toward the next generation of recommender systems: a survey of the state-of-the-art and possible extensions. IEEE Trans. Knowl. Data Eng. **17**(6), 734–749 (2005). https://doi.org/10.1109/TKDE.2005.99
2. Afoudi, Y., Lazaar, M., Al Achhab, M.: Hybrid recommendation system combined content-based filtering and collaborative prediction using artificial neural network. Simul. Model. Pract. Theory **113**, 102375 (2021). https://doi.org/10.1016/j.simpat.2021.102375

3. Aggarwal, C.C.: Ensemble-Based and Hybrid Recommender Systems, pp. 199–224. Springer International Publishing, Cham (2016). https://doi.org/10.1007/978-3-319-29659-3_6

4. Agrawal, R., Srikant, R.: Fast algorithms for mining association rules in large databases. In: Proceedings of the 20th International Conference on Very Large Data Bases, VLDB '94, pp. 487–499. Morgan Kaufmann Publishers Inc., San Francisco (1994)

5. Anelli, V.W., Bellogin, A., Ferrara, A., Malitesta, D., Merra, F.A., Pomo, C., Donini, F.M., Di Noia, T.: Elliot: A comprehensive and rigorous framework for reproducible recommender systems evaluation. In: Proceedings of the 44th International ACM SIGIR Conference on Research and Development in Information Retrieval, SIGIR'21, pp. 2405–2414. Association for Computing Machinery, New York (2021). https://doi.org/10.1145/3404835.3463245

6. Bai, X., Wang, M., Lee, I., Yang, Z., Kong, X., Xia, F.: Scientific paper recommendation: a survey. IEEE Access **7**, 9324–9339 (2019). https://doi.org/10.1109/ACCESS.2018.2890388

7. Balabanović, M., Shoham, Y.: Fab: content-based, collaborative recommendation. Commun. ACM **40**(3), 66–72 (1997). https://doi.org/10.1145/245108.245124

8. Bendouch, M.M., Frasincar, F., Robal, T.: Addressing scalability issues in semantics-driven recommender systems. In: IEEE/WIC/ACM International Conference on Web Intelligence and Intelligent Agent Technology, WI-IAT '21, pp. 56–63. Association for Computing Machinery, New York (2021). https://doi.org/10.1145/3486622.3493963

9. Breese, J.S., Heckerman, D., Kadie, C.: Empirical analysis of predictive algorithms for collaborative filtering. In: Proceedings of the Fourteenth Conference on Uncertainty in Artificial Intelligence, UAI'98, pp. 43–52. Morgan Kaufmann Publishers Inc., San Francisco (1998)

10. Burke, R.: Hybrid recommender systems: Survey and experiments. User Modeling and User-Adapted Interaction **12** (2002). https://doi.org/10.1023/A:1021240730564

11. Chen, R., Hua, Q., Chang, Y.S., Wang, B., Zhang, L., Kong, X.: A survey of collaborative filtering-based recommender systems: from traditional methods to hybrid methods based on social networks. IEEE Access **6**, 64301–64320 (2018). https://doi.org/10.1109/ACCESS.2018.2877208

12. Cunha, T., Soares, C., de Carvalho, A.C.: Metalearning and recommender systems: a literature review and empirical study on the algorithm selection problem for collaborative filtering. Inf. Sci. **423**, 128–144 (2018). https://doi.org/10.1016/j.ins.2017.09.050

13. Ekstrand, M.D., Riedl, J.T., Konstan, J.A.: Collaborative filtering recommender systems. Found. Trends Hum.-Comput. Interact. **4**(2), 81–173 (2011). https://doi.org/10.1561/1100000009

14. Fayyad, U.: Knowledge Discovery in Databases: An Overview, pp. 28–47. Springer, Berlin (2001). https://doi.org/10.1007/978-3-662-04599-2_2

15. Feng, J., Xia, Z., Feng, X., Peng, J.: RBPR: a hybrid model for the new user cold start problem in recommender systems. Knowl.-Based Syst. **214**, 106732 (2021). https://doi.org/10.1016/j.knosys.2020.106732

16. Fouss, F., Fernandes, E.: A closer-to-reality model for comparing relevant dimensions of recommender systems, with application to novelty. Information **12**(12) (2021). https://www.mdpi.com/2078-2489/12/12/500

17. Freudenthaler, C., Schmidt-thieme, L., Rendle, S.: Bayesian factorization machines (2010)

18. Freudenthaler, C., Schmidt-Thieme, L., Rendle, S.: Factorization machines factorized polynomial regression models (2011)

19. de Gemmis, M., Lops, P., Musto, C., Narducci, F., Semeraro, G.: Semantics-Aware Content-Based Recommender Systems, pp. 119–159. Springer US, Boston (2015). https://doi.org/10.1007/978-1-4899-7637-6_4

20. Grzegorowski, M., Litwin, J., Wnuk, M., Pabis, M., Marcinowski, L.: Survival-based feature extraction - application in supply management for dispersed vending machines. IEEE Trans. Industr. Inf. (2022). https://doi.org/10.1109/TII.2022.3178547

21. Grzegorowski, M., Zdravevski, E., Janusz, A., Lameski, P., Apanowicz, C., Ślęzak, D.: Cost optimization for big data workloads based on dynamic scheduling and cluster-size tuning. Big Data Res. **25**, 100203 (2021). https://doi.org/10.1016/j.bdr.2021.100203

22. Gunawardana, A., Shani, G.: Evaluating Recommender Systems, pp. 265–308. Springer US, Boston (2015). https://doi.org/10.1007/978-1-4899-7637-6_8
23. Han, J., Kamber, M., Pei, J.: 6 - mining frequent patterns, associations, and correlations: basic concepts and methods. In: Han, J., Kamber, M., Pei, J. (eds.), Data Mining (Third Edition), The Morgan Kaufmann Series in Data Management Systems, 3rd edn., pp. 243–278. Morgan Kaufmann, Boston (2012). https://www.sciencedirect.com/science/article/pii/B978012381479100006X
24. Han, J., Pei, J., Yin, Y.: Mining frequent patterns without candidate generation. SIGMOD Rec. 29(2), 1–12 (2000). https://doi.org/10.1145/335191.335372
25. Himabindu, T., Padmanabhan, V., Pujari, A.K.: Conformal matrix factorization based recommender system. Inf. Sci. 467 (2018). https://doi.org/10.1016/j.ins.2018.04.004
26. Himeur, Y., Alsalemi, A., Al-Kababji, A., Bensaali, F., Amira, A., Sardianos, C., Dimitrakopoulos, G., Varlamis, I.: A survey of recommender systems for energy efficiency in buildings: Principles, challenges and prospects. Inf. Fus. 72, 1–21 (2021). https://doi.org/10.1016/j.inffus.2021.02.002
27. Hipp, J., Güntzer, U., Nakhaeizadeh, G.: Algorithms for association rule mining - a general survey and comparison. SIGKDD Explor. Newsl. 2(1), 58–64 (2000). https://doi.org/10.1145/360402.360421
28. Idrissi, N., Zellou, A.: A systematic literature review of sparsity issues in recommender systems. Soc. Netw. Anal. Min. 10(1), 15 (2020). https://doi.org/10.1007/s13278-020-0626-2
29. Jalili, M., Ahmadian, S., Izadi, M., Moradi, P., Salehi, M.: Evaluating collaborative filtering recommender algorithms: a survey. IEEE Access 6, 74003–74024 (2018). https://doi.org/10.1109/ACCESS.2018.2883742
30. Kannout, E.: Context clustering-based recommender systems. In: 2020 15th Conference on Computer Science and Information Systems (FedCSIS), vol. 21, pp. 85–91 (2020). https://doi.org/10.15439/2020F54
31. Karimi, M., Jannach, D., Jugovac, M.: News recommender systems - survey and roads ahead. Inf. Process. Manag. 54(6), 1203–1227 (2018). https://doi.org/10.1016/j.ipm.2018.04.008
32. Ko, H., Lee, S., Park, Y., Choi, A.: A survey of recommendation systems: recommendation models, techniques, and application fields. Electronics 11(1) (2022). https://www.mdpi.com/2079-9292/11/1/141
33. Koren, Y., Bell, R.: Advances in Collaborative Filtering, pp. 77–118. Springer US, Boston (2015). https://doi.org/10.1007/978-1-4899-7637-6_3
34. Lika, B., Kolomvatsos, K., Hadjiefthymiades, S.: Facing the cold start problem in recommender systems. Expert Syst. Appl. 41(4, Part 2), 2065 2073 (2014). https://doi.org/10.1016/j.eswa.2013.09.005
35. Lin, C.J., Kuo, T.T., Lin, S.D.: A content-based matrix factorization model for recipe recommendation. In: Tseng, V.S., Ho, T.B., Zhou, Z.H., Chen, A.L.P., Kao, H.Y. (eds.) Advances in Knowledge Discovery and Data Mining, pp. 560–571. Springer International Publishing, Cham (2014)
36. Miyahara, K., Pazzani, M.: Collaborative filtering with the simple Bayesian classifier. Inf. Process. Soc. Japan 43 (2002). https://doi.org/10.1007/3-540-44533-1_68
37. Mohamed, M.H., Khafagy, M.H., Ibrahim, M.H.: Recommender systems challenges and solutions survey. In: 2019 International Conference on Innovative Trends in Computer Engineering (ITCE), pp. 149–155 (2019). https://doi.org/10.1109/ITCE.2019.8646645
38. Movahedian, H., Khayyambashi, M.R.: Folksonomy-based user interest and disinterest profiling for improved recommendations: an ontological approach. J. Inf. Sci. 40(5), 594–610 (2014). https://doi.org/10.1177/0165551514539870
39. Navgaran, D.Z., Moradi, P., Akhlaghian, F.: Evolutionary based matrix factorization method for collaborative filtering systems. In: 2013 21st Iranian Conference on Electrical Engineering (ICEE), pp. 1–5 (2013). https://doi.org/10.1109/IranianCEE.2013.6599844
40. Nguyen, H.S.: Efficient machine learning methods over pairwise space (keynote). In: Schlingloff, H., Vogel, T., (eds.) Proceedings of the 29th International Workshop on Concurrency, Specification and Programming (CS&P 2021), Berlin, Germany, September 27–28, 2021,

CEUR Workshop Proceedings, vol. 2951, pp. 117–119. CEUR-WS.org (2021). http://ceur-ws.org/Vol-2951/keynote2.pdf

41. Park, M.H., Hong, J.H., Cho, S.B.: Location-based recommendation system using Bayesian user's preference model in mobile devices. In: Indulska, J., Ma, J., Yang, L.T., Ungerer, T., Cao, J. (eds.) Ubiquit. Intell. Comput., pp. 1130–1139. Springer, Berlin (2007)
42. Pawlicka, A., Pawlicki, M., Kozik, R., Choraś, R.S.: A systematic review of recommender systems and their applications in cybersecurity. Sensors 21(15) (2021). https://www.mdpi.com/1424-8220/21/15/5248
43. Pazzani, M.J., Billsus, D.: Content-Based Recommendation Systems, pp. 325–341. Springer, Berlin (2007). https://doi.org/10.1007/978-3-540-72079-9_10
44. Pérez-Almaguer, Y., Yera, R., Alzahrani, A.A., Martínez, L.: Content-based group recommender systems: a general taxonomy and further improvements. Expert Syst. Appl. 184, 115444 (2021). https://doi.org/10.1016/j.eswa.2021.115444
45. Pérez-Almaguer, Y., Yera, R., Alzahrani, A.A., Martínez, L.: Content-based group recommender systems: a general taxonomy and further improvements. Expert Syst. Appl. 184, 115444 (2021). https://doi.org/10.1016/j.eswa.2021.115444. www.sciencedirect.com/science/article/pii/S0957417421008587
46. Philip, S., Shola, P.B., John, A.O.: Application of content-based approach in research paper recommendation system for a digital library. Int. J. Adv. Comput. Sci. Appl. 5 (2014)
47. Porcel, C., López-Herrera, A., Herrera-Viedma, E.: A recommender system for research resources based on fuzzy linguistic modeling. Expert Syst. Appl. 36(3, Part 1), 5173–5183 (2009). https://doi.org/10.1016/j.eswa.2008.06.038. https://www.sciencedirect.com/science/article/pii/S0957417408003126
48. Quadrana, M., Cremonesi, P., Jannach, D.: Sequence-aware recommender systems. ACM Comput. Surv. 51(4), 1–36 (2019). https://doi.org/10.1145/3190616
49. Ranjbar, M., Moradi, P., Azami, M., Jalili, M.: An imputation-based matrix factorization method for improving accuracy of collaborative filtering systems. Eng. Appl. Artif. Intell. 46, 58–66 (2015). https://doi.org/10.1016/j.engappai.2015.08.010. www.sciencedirect.com/science/article/pii/S0952197615001888
50. Rendle, S.: Factorization machines. In: 2010 IEEE International Conference on Data Mining, pp. 995–1000 (2010). https://doi.org/10.1109/ICDM.2010.127
51. Rendle, S.: Factorization machines with libfm. ACM Trans. Intell. Syst. Technol. 3(3) (2012). https://doi.org/10.1145/2168752.2168771
52. Rendle, S., Gantner, Z., Freudenthaler, C., Schmidt-Thieme, L.: Fast context-aware recommendations with factorization machines. In: Proceedings of the 34th International ACM SIGIR Conference on Research and Development in Information Retrieval, SIGIR '11, pp. 635–644. Association for Computing Machinery, New York (2011). https://doi.org/10.1145/2009916.2010002
53. Ricci, F., Rokach, L., Shapira, B.: Recommender Systems: Introduction and Challenges, pp. 1–34. Springer US, Boston (2015). https://doi.org/10.1007/978-1-4899-7637-6_1
54. Sarwar, B., Karypis, G., Konstan, J., Riedl, J.: Application of dimensionality reduction in recommender system – a case study. In: ACM WebKDD'00 (Web-mining for ECommerce Workshop (2000). https://doi.org/10.21236/ada439541
55. Silveira, T., Zhang, M., Lin, X., Liu, Y., Ma, S.: How good your recommender system is? A survey on evaluations in recommendation. Int. J. Mach. Learn. Cybern. 10(5), 813–831 (2019). https://doi.org/10.1007/s13042-017-0762-9
56. Singh, M.: Scalability and sparsity issues in recommender datasets: a survey. Knowl. Inf. Syst. 62(1), 1–43 (2020). https://doi.org/10.1007/s10115-018-1254-2
57. Terán, L., Meier, A.: A fuzzy recommender system for elections. In: Andersen, K.N., Francesconi, E., Grönlund, Å., van Engers, T.M. (eds.) Electronic Government and the Information Systems Perspective, pp. 62–76. Springer, Berlin (2010)
58. Thorat, P.B., Goudar, R.M., Barve, S.: Article: Survey on collaborative filtering, content-based filtering and hybrid recommendation system. Int. J. Comput. Appl. 110(4), 31–36 (2015)

59. Tilahun, Z., Jun, H., Oad, A.: Solving cold-start problem by combining personality traits and demographic attributes in a user based recommender system. Int. J. Adv. Res. Comput. Sci. Softw. Eng. **7**(5), 231–239 (2017). https://doi.org/10.23956/ijarcsse/v7i4/01420

60. Vargas, S., Castells, P.: Rank and relevance in novelty and diversity metrics for recommender systems. In: Proceedings of the Fifth ACM Conference on Recommender Systems, RecSys '11, pp. 109–116. Association for Computing Machinery, New York (2011). https://doi.org/10.1145/2043932.2043955

61. Walek, B., Fojtik, V.: A hybrid recommender system for recommending relevant movies using an expert system. Expert Syst. Appl. **158**, 113452 (2020). https://doi.org/10.1016/j.eswa.2020.113452

62. Wang, D., Liang, Y., Xu, D., Feng, X., Guan, R.: A content-based recommender system for computer science publications. Knowl.-Based Syst. **157**, 1–9 (2018). https://doi.org/10.1016/j.knosys.2018.05.001. www.sciencedirect.com/science/article/pii/S0950705118302107

63. Wu, Z., Li, C., Cao, J., Ge, Y.: On scalability of association-rule-based recommendation: a unified distributed-computing framework. ACM Trans. Web **14**(3) (2020). https://doi.org/10.1145/3398202

64. Ying, R., He, R., Chen, K., Eksombatchai, P., Hamilton, W.L., Leskovec, J.: Graph convolutional neural networks for web-scale recommender systems. In: Proceedings of the 24th ACM SIGKDD, KDD '18, pp. 974–983. Association for Computing Machinery, New York (2018). https://doi.org/10.1145/3219819.3219890

65. Zhou, T., Kuscsik, Z., Liu, J.G., Medo, M., Wakeling, J., Zhang, Y.C.: Solving the apparent diversity-accuracy dilemma of recommender systems. Proc. Natl. Acad. Sci. U.S.A. **107**, 4511–5 (2010). https://doi.org/10.1073/pnas.1000488107

66. Ziegler, C.N., McNee, S.M., Konstan, J.A., Lausen, G.: Improving recommendation lists through topic diversification. In: Proceedings of the 14th international conference on World Wide Web, WWW '05. Association for Computing Machinery, New York (2005). https://doi.org/10.1145/1060745.1060754

Chapter 6
Security Enforcing

Damas P. Gruska

Abstract Formal definitions of state-based and language-based security with respect to timing attacks are proposed and studied. Then various ways how to secure systems with respect to such attacks are discussed. First, we investigate time insertion functions. Conditions, when such functions exist and could protect systems, are investigated. Then we discuss the concept of supervisor control which can be used if there is no appropriate time insertion function to protect the systems.

6.1 Introduction

Exploiting formal models and methods to define, study, or even discover system vulnerability regarding security threats is a hot topic nowadays. Formal methods allow us, in many cases, to show, even prove, that a given system is not secure. Then we have a couple of options on what to do. We can either re-design its behavior, which might be costly, difficult, or even impossible, in the case that it is already part of a hardware solution, proprietary firmware, and so on, or we can use supervisory control (see [27, 28]) to restrict the system's behavior in such a way that the system becomes secure. A supervisor can see (some) system's actions and can control (disable or enable) some set of system actions. In this way, it restricts the system's behavior to guarantee its security (see also [14]). This is a trade-off between security and functionality. The situation is different in the case of timing attacks. They use timing information leakage, which is an ability for an attacker to deduce internal (private) information depending on timing information. They, as side-channel attacks, represent a serious threat to many systems. They allow intruders "break" "unbreakable" systems, algorithms, protocols, etc. For example, by carefully measuring the amount of time required to perform private key operations, attackers may be able to find fixed Diffie-Hellman exponents, factor RSA keys, and break other cryptosystems (see [23]). This idea was developed in [5] where a timing attack against smart card

D. P. Gruska (✉)
Department of Applied Informatics, Comenius University, Bratislava, Slovak Republic
e-mail: gruska@fmph.uniba.sk

© The Author(s), under exclusive license to Springer Nature Switzerland AG 2023
B. Schlingloff et al. (eds.), *Concurrency, Specification and Programming, Studies in Computational Intelligence 1091*, https://doi.org/10.1007/978-3-031-26651-5_6

implementation of RSA was conducted. In [16], a timing attack on the RC5 block encryption algorithm is described. The analysis is motivated by the possibility that some implementations of RC5 could result in data-dependent rotations taking a time that is a function of the data. In [17], the vulnerability of two implementations of the Data Encryption Standard (DES) cryptosystem under a timing attack is studied. It is shown that a timing attack yields the Hamming weight of the key used by both DES implementations. Moreover, the attack is computationally inexpensive. A timing attack against an implementation of AES candidate Rijndael is described in [24], and the one against the popular SSH protocol in [30]. In [2] several novel timing attacks against the common table-driven software implementation of the AES cipher are described. Also possible attacks on most of the currently used processors (Meltdown and Spectre) belong to timing attacks. Timing attacks on web privacy and some corresponding formal models can be found in [7].

In this paper, we first formalize security as the absence of information flow between private and public activities or data. We exploit concept opacity. Opacity states that if there exists a run of the system that reveals the secret (some property of actions or states) there exists another run, with the same observation, that does not reveal that secret. In the literature, we can find various types and various ways how to deal with information flow based attacks (see, for example, [7, 25, 29]). Here we concentrate on timing attacks. Much research is devoted to the study of specific types of timing attacks, where the attacker sees the initial and final state of the system and only the time that passes between them. There is a specific intermediate private state and intruder tries to deduce whether the system went through this state. If it cannot be deduced then the system is opaque (see [1] for formalization of this property in a framework of Parametric Timed Automata). Here we will work with different security properties. We expect an intruder who sees not only the passing of time, and instead of an intermediate private state, we work with properties of final states or sequences of actions that lead to them. If the system is not secure due to timing attacks, we suggest protecting it first using insertion functions (see [18, 19, 22, 33] or [21] for general use of insertion functions). Such functions can add some idling between actions to enforce security. We investigate conditions under which such functions do exist and also their properties. Later we propose to use a supervisor controller as a way to protect systems in the case that there does not exist an appropriate time insertion function. As for the formalism, we will work with a timed process algebra and state-based and language-based opacity. This formalism allows us to formalize timing attacks. Process algebras unlike finite automata or Timed Automata, allows us to use various compositional constructs, to express observations as well as intruders themselves. In [12] we have introduced time insertion functions to guarantee language-based security and showed some of their properties (recalled here by Definitions 6.6, 6.11 and corresponding propositions). In [12, 13] we studied conditions under which there exists a timed insertion function for a given process and security language-based security property and we have presented some decidability and undecidability results. In [15] we have studied state-based security with respect to timing attacks and corresponding time insertion functions (recalled here by Definitions 6.8, 6.12 and corresponding propositions. In [14] we study supervisor controllers as an alternative way to protect systems. Here (Sect. 6.4), we recall this concept with some modifications and suggest how it can be used in combination with timed insertion functions. Hence this paper synthesizes and integrates all above mentioned previous works on insertion functions and supervisory control.

Despite the fact that this work is primarily devoted to theoretical research, as a large body of research in this area, it also has application value. To avoid a possible timing side-channel attack, as described for example in [23], in practice, time insertion functions are used that insert random delays. It is worth studying how powerful and when applicable could these functions be. The obtained results could be directly applicable to programming languages based on process algebras or using labeled transition systems to any imperative programming language (as stated in [1], where translation (Parametric Timed Automata—labeled transition system—a set of Java programs) is investigated.

The paper is organized as follows. In Sect. 6.2 we describe the timed process algebra TPA which will be used as a basic formalism and state-based and language-based security properties. The next section is devoted to time insertion functions as a way how to guarantee security with respect to timing attacks. Section 6.4 is devoted to supervisory control and how it can be exploited in case there is no time insertion function that can protect systems against timing attacks.

6.2 Working Formalism

In this section, we briefly recall our working formalism which will be based on Timed Process Algebra, TPA for short. TPA itself is based on Milner's Calculus of Communicating Systems (for short, CCS, see [26]), so we will start with this. To define the language CCS, we first presuppose a set of atomic action symbols A not containing symbols τ, and that for every $a \in A$ there exists an $\bar{a} \in A$ and $\bar{\bar{a}} = a$ i.e. actions which represent receiving and sending from a channel a, respectively. We define $Act = A \cup \{\tau\}$ where τ represent internal action, for example, a result of internal communication. We let $a, b, \ldots$ range over $A; x, y, \ldots$ range over Act. The set of CSS terms is defined by the following BNF expression where $x \in Act$, $X \in Var$, Var is a set of process variables, S is ranging over relabelling functions, $S : Act \to Act$ is such that $\overline{S(a)} = S(\bar{a})$ for $a \in A$, $S(\tau) = \tau$ and finally, $L \subseteq A$:

$$P ::= Nil \mid X \mid x.P \mid P + P \mid P \mid P \mid \mu X P \mid P \setminus L \mid P[S]$$

We use the usual definitions for free and bound variables, open and closed terms and guarded recursion. The set **CCS** of *processes* consists of closed and guarded CCS terms.

Nil represents process doing nothing, X is a process variable. Process $x.P$ can perform action x and then behaves as process P (prefix operator) , $+,|$ represent nondeterministic choice and parallel operator, respectively. A recursion operator is denoted by $\mu X P$ i.e. recursive definition of process given by equation $X = P$. Unary operators $P \setminus L$ and $P[S]$ represent restriction and relabeling, respectively. Formal definition of a structural operational semantics for **CCS** is defined in terms of Labeled Transition Systems.

Definition 6.1 *A Labeled Transition System* is a triple $(S, T, \{\overset{x}{\rightarrow}, x \in T\})$ where S is a set of states, T is a set of labels and $\{\overset{x}{\rightarrow}, x \in T\}$ is the transition relation such that each $\overset{x}{\rightarrow}$ is a binary transition relation on S. We write $P \overset{x}{\rightarrow} P'$ instead of $(P, P') \in \overset{x}{\rightarrow}$ and $P \overset{x}{\not\rightarrow}$ if there is no P' such that $P \overset{x}{\rightarrow} P'$.

As a set of states we use set of CCS terms, set of labels is equal to Act and transition relations are defined as follows.

Definition 6.2 The transition relations $T = \{\overset{x}{\rightarrow}_{CCS}, x \in Act\}$ are defined as the least relations satisfying the following inference rules:

$$\frac{}{x.P \overset{x}{\rightarrow} P} \qquad\qquad \frac{P \overset{x}{\rightarrow} P'}{P + Q \overset{x}{\rightarrow} P', \; Q + P \overset{x}{\rightarrow} P'}$$

$$\frac{P \overset{x}{\rightarrow} P'}{P \setminus L \overset{x}{\rightarrow} P' \setminus L}, (x, \bar{x} \notin L) \qquad\qquad \frac{P[\mu X P / X] \overset{x}{\rightarrow} P'}{\mu X P \overset{x}{\rightarrow} P'}$$

$$\frac{P \overset{x}{\rightarrow} P'}{P[S] \overset{S(x)}{\rightarrow} P'[S]} \qquad\qquad \frac{P \overset{u}{\rightarrow} P', Q \overset{\bar{a}}{\rightarrow} Q'}{P \mid Q \overset{\tau}{\rightarrow} P' \mid Q'}$$

$$\frac{P \overset{u}{\rightarrow} P'}{P \mid Q \overset{u}{\rightarrow} P' \mid Q, Q \mid P \overset{u}{\rightarrow} Q \mid P'}$$

We are now ready to define the working formalism, which is the time extension of CCS. In TPA we use the special time action t which expresses elapsing of (discrete) time is added and hence the set of actions is extended from Act to $Actt$. The presented language is a slight simplification of Timed Security Process Algebra (tSPA) introduced in [8]. We omit an explicit idling operator ι used in tSPA and instead of this we allow implicit idling of processes. Hence processes can perform either "enforced idling" by performing t actions which are explicitly expressed in their descriptions or "voluntary idling" (i.e. for example, the process $a.Nil$ can perform t action despite the fact that this action is not formally expressed in the process specification). But in both cases, internal communications have priority to action t in the parallel composition. Moreover, we do not divide actions into private and public ones as it is in tSPA. TPA differs also from the tCryptoSPA (see [9]). TPA does not use value passing and strictly preserves *time determinacy* in case of choice operator $+$ what is not the case of tCryptoSPA (see [10]). To define $At = A \cup \{t\}$, $Actt = Act \cup \{t\}$, moreover we suppose that $S(t) = t$ for every relabelling function S. We give a structural operational semantics of terms again by means of labeled transition systems. The set of terms represents a set of states, labels are actions from $Actt$. The transition relation $\rightarrow$ is a subset of TPA $\times$ $Actt$ $\times$ TPA. We define the transition relation as the least relation satisfying the inference rules for CCS plus the following inference rules for t action (for more details see [10]).

$$\frac{}{Nil \overset{t}{\rightarrow} Nil} \quad A1 \qquad\qquad \frac{}{u.P \overset{t}{\rightarrow} u.P} \quad A2$$

$$\frac{P \overset{t}{\rightarrow} P', Q \overset{t}{\rightarrow} Q', P \mid Q \overset{\tau}{\nrightarrow}}{P \mid Q \overset{t}{\rightarrow} P' \mid Q'} \quad Pa \qquad \frac{P \overset{t}{\rightarrow} P', Q \overset{t}{\rightarrow} Q'}{P + Q \overset{t}{\rightarrow} P' + Q'} \quad S$$

For $s = x_1.x_2.\ldots.x_n, x_i \in Act$ we write $P \overset{s}{\rightarrow}$ instead of $P \overset{x_1}{\rightarrow} \overset{x_2}{\rightarrow} \cdots \overset{x_n}{\rightarrow}$ and we say that s is a trace of P. The set of all traces of P will be denoted by $Tr(P)$. By ϵ we denote the empty sequence and by M^* we denote the set of finite sequences of elements from M. We use $\overset{x}{\Rightarrow}$ as an abbreviation for transitions including τ actions i.e. $(\overset{\tau}{\rightarrow})^* \overset{x}{\rightarrow} (\overset{\tau}{\rightarrow})^*$ (see [26]). By $s|_B$ we will denote the sequence obtained from s by removing all actions not belonging to B. By $L(P)$ we will denote a set of actions which can be performed by P, i.e. $L(P) = \{x \mid P \overset{s.x}{\rightarrow}, s \in Actt^*\}$. We define $\widehat{x} = x$ if $x \neq \tau$ and $\widehat{x} = \epsilon$ otherwise.

Now we define two behavior equivalences trace equivalence and weak bisimulation, respectively (see [26]).

Definition 6.3 The set of weak traces of process P is defined as $Tr_w(P) = \{s \in At^* \mid \exists P'.P \overset{s}{\Rightarrow} P'\}$.

Two processes P and Q are weakly trace equivalent iff $Tr_w(P) = Tr_w(Q)$.

Definition 6.4 Let (TPA, Act, $\rightarrow$) be a labeled transition system (LTS). A relation $\Re \subseteq$ TPA $\times$ TPA is called a *weak bisimulation* if it is symmetric and it satisfies the following condition: if $(P, Q) \in \Re$ and $P \overset{x}{\rightarrow} P', x \in Actt$ then there exists a process Q' such that $Q \overset{\widehat{x}}{\Rightarrow} Q'$ and $(P', Q') \in \Re$. Two processes P, Q are *weakly bisimilar*, abbreviated $P \approx Q$, if there exists a weak bisimulation relating P and Q.

To formalize an information flow we do not divide actions into public and private ones at the system description level, as it is done for example in [9], but we use a more general concept of observation and opacity. This concept was exploited in [3, 4] in a framework of Petri Nets and transition systems, respectively. Firstly we define the observation function on sequences from Act^*. Various variants of observation functions differ according to contexts which they take into account. For example, an observation of action can depend on the previous actions.

Definition 6.5 (*Observation*) Let Θ be a set of elements called observables. Any function $\mathcal{O} : Actt^* \rightarrow \Theta^*$ is an observation function. It is called static/dynamic/orwellian/m-orwellian $(m \geq 1)$ if   the   following   conditions   hold respectively (below we assume $w = x_1 \ldots x_n$):

- static if there is a mapping $\mathcal{O}' : Actt \rightarrow \Theta \cup \{\epsilon\}$ such that for every $w \in Act^\star$ it holds $\mathcal{O}(w) = \mathcal{O}'(x_1) \dots \mathcal{O}'(x_n)$,
- dynamic if there is a mapping $\mathcal{O}' : Actt^\star \rightarrow \Theta \cup \{\epsilon\}$ such that for every $w \in Actt^\star$ it holds $\mathcal{O}(w) = \mathcal{O}'(x_1).\mathcal{O}'(x_1.x_2) \dots \mathcal{O}'(x_1 \dots x_n)$,
- orwellian if there is a mapping $\mathcal{O}' : Actt \times Actt^\star \rightarrow \Theta \cup \{\epsilon\}$ such that for every $w \in Actt^\star$ it holds $\mathcal{O}(w) = \mathcal{O}'(x_1, w).\mathcal{O}'(x_2, w) \dots \mathcal{O}'(x_n, w)$,
- m-orwellian if there is a mapping $\mathcal{O}' : Actt \times Actt^\star \rightarrow \Theta \cup \{\epsilon\}$ such that for every $w \in Actt^\star$ it holds $\mathcal{O}(w) = \mathcal{O}'(x_1, w_1).\mathcal{O}'(x_2, w_2) \dots \mathcal{O}'(x_n, w_n)$ where $w_i = x_{max\{1, i-m+1\}}.x_{max\{1, i-m+1\}+1} \cdots x_{min\{n, i+m-1\}}$.

In the case of the static observation function, each action is observed independently from its context. In the case of the dynamic observation function, an observation of an action depends on the previous ones, in the case of the orwellian and m-orwellian observation function an observation of an action depends on the all and on m previous actions in the sequence, respectively. The static observation function is the special case of m-orwellian one for $m = 1$. Note that from the practical point of view the m-orwellian observation functions are the most interesting ones. An observation expresses what an observer—eavesdropper can see from a system behavior and we will alternatively use both the terms (observation—observer) with the same meaning. Note that the same action can be seen differently during an observation (except static observation function) and this expresses a possibility to accumulate some knowledge by an intruder. For example, an action not visible at the beginning could become somehow observable. An observation function can be naturally extended to set of sequences. From now on we will assume that $\Theta \subseteq Actt$.

Now suppose that we have some security property over process traces. This might be an execution of one or more classified actions, an execution of actions in a particular classified order which should be kept hidden, etc. Suppose that this property is expressed by the predicate ϕ over process traces. We would like to know whether an observer can deduce the validity of the property ϕ just by observing sequences of actions from $Actt^\star$ performed by given process. The observer cannot deduce the validity of ϕ if there are two traces $w, w' \in Actt^\star$ such that $\phi(w), \neg\phi(w')$ and the traces cannot be distinguished by the observer i.e. $\mathcal{O}(w) = \mathcal{O}(w')$. We formalize this concept by opacity.

Definition 6.6 (*Opacity*) Given process P, a predicate ϕ over $Actt^\star$ is opaque w.r.t. the observation function $\mathcal{O}$ if for every sequence $w, w \in Tr(P)$ such that $\phi(w)$ holds and $\mathcal{O}(w) \neq \epsilon$, there exists a sequence $w', w' \in Tr(P)$ such that $\neg\phi(w')$ holds and $\mathcal{O}(w) = \mathcal{O}(w')$. The set of processes for which the predicate ϕ is opaque with respect to $\mathcal{O}$ will be denoted by $Op_{\mathcal{O}}^{\phi}$.

A predicate is opaque if for any trace of a system for which it holds, there exists another trace for which it does not hold and both traces are indistinguishable for an observer (which is expressed by an observation function). This means that the observer (intruder) cannot say whether a trace for which the predicate holds has been performed or not.

Suppose that all actions are divided into two groups, namely public (low level) actions L and private (high level) actions H. It is assumed that $L \cup H = A$. Strong Nondeterministic Non-Interference (SNNI, for short, see [8]) property assumes an intruder who tries to learn whether a private action was performed by a given process while (s)he can observe only public ones. If this cannot be done, then the process has SNNI property. Note that SNNI property is the special case of opacity for static observation function $\mathcal{O}(x) = x$ iff $x \notin H$ and $\mathcal{O}(x) = \epsilon$ otherwise, and $\phi(w)$ is such that it is true iff w contains an action from H.

Now let us assume that an intruder is interested in whether a given process has reached a state with some given property which is expressed by a (total) predicate. This property might be process deadlock, capability to execute only traces with some given actions, capability to perform the same actions from a given set, incapacity to idle (to perform τ action), etc. We do not put any restrictions on such predicates but we only assume that they are consistent with some suitable behavioral equivalence. The formal definition follows.

Definition 6.7 We say that the predicate ϕ over processes is consistent with respect to relation $\cong$ if whenever $P \cong P'$ then $\phi(P) \Leftrightarrow \phi(P')$.

As the consistency relation $\cong$ we could take bisimulation, weak bisimulation, weak trace equivalence or any other suitable equivalence.

An intruder cannot learn validity of predicate ϕ observing process's behaviour iff there are two traces, undistinguished for him (by observation function $\mathcal{O}$), where one leads to a state which satisfy ϕ and another one leads to a state for which $\neg\phi$ holds. The formal definition follows.

Definition 6.8 (*Process Opacity*) Given process P, a predicate ϕ over processes is process opaque w.r.t. the observation function $\mathcal{O}$ whenever $P \xrightarrow{w} P'$ for $w \in Actt^*$ and $\phi(P')$ holds then there exists P'' such that $P \xrightarrow{w'} P''$ for some $w' \in Actt^*$ and $\neg\phi(P'')$ holds and moreover $\mathcal{O}(w) = \mathcal{O}(w')$. The set of processes for which the predicate ϕ is process opaque w.r.t. to the $\mathcal{O}$ will be denoted by $POp_{\mathcal{O}}^{\phi}$.

Opacity is language-based security and process opacity is a state-based security property, we will refer to properties related to them as L and S properties, respectively.

6.3 Insertion Functions

Timing attacks belong to powerful tools for attackers who can observe or interfere with systems in real time. On the other side these techniques are useless for off-line systems and hence they could be consider safe with respect to attackers who cannot observe (real) time behaviour. By the presented formalism we have a way how to distinguish these two cases. First we define untimed version of an observation function, i.e. a function which does not take into account time information. From now on we will consider observation functions $\mathcal{O} : Actt^* \to Actt^*$ for which there exists untimed variants $\mathcal{O}_t$. Function $\mathcal{O}_t$ is untimed variant of $\mathcal{O}$ iff $\mathcal{O}_t(w) = \mathcal{O}(w)|_{Act}$, i.e. untimed variant represents an observer who does not see elapsing of time. Now we can formally describe situation when a process could be jeopardized by a timing attack i.e. is secure only if an observer cannot see elapsing of time.

Definition 6.9 (*Timinig Attacks*) We say that process P is prone to S-timing attack (L-timing attack) with respect to ϕ and $\mathcal{O}$ iff $P \notin POp_{\mathcal{O}}^{\phi}$ but $P \in POp_{\mathcal{O}_t}^{\phi}$ ($P \notin Op_{\mathcal{O}}^{\phi}$ but $P \in Op_{\mathcal{O}_t}^{\phi}$).

Note 6.1 Let us assume an intruder who tries to learn whether a a state which satisfy ϕ is reached. Then process $P = a.t.R + b.t.t.Q$ is not opaque for static observation function $\mathcal{O}(x) = \epsilon$ for $x \neq t$, $\mathcal{O}(t) = t$ and $\phi(R)$, $\neg\phi(Q)$ hold, i.e. $P \notin POp_{\mathcal{O}}^{\phi}$. But if an observer cannot see elapsing of time this process is opaque, i.e. $P \in POp_{\mathcal{O}_t}^{\phi}$. Hence P is prone to S-timing attack.

Note that in [1] a different type of timing leakage is studied. An observer can see only elapsing of time i.e. $\mathcal{O}(t) = t$ and $\mathcal{O}(x) = \epsilon$ for $x \neq t$. System is opaque if for an given execution time d there exist two runs of duration d from initial to finial states , one going the through private state and another one not going through the private state.

Note 6.2 Let us assume an intruder who tries to learn whether a private action h was performed by a given process while (s)he can observe only public action l but not h itself. Then process $P = t.t.t.l.Nil + t.h.t.Nil$ is not opaque for static observation function $\mathcal{O}(x) = x$ iff $x \in \{l, t\}$, $\mathcal{O}(h) = \epsilon$ and $\phi(w)$ is such that it is true iff w contains an action h, i.e. $P \notin Op_{\mathcal{O}}^{\phi}$. But if an observer cannot see elapsing of time this process is opaque, i.e. $P \in Op_{\mathcal{O}_t}^{\phi}$. Hence P is prone to L-timing attack.

From now on we will consider only processes which are prone to timing attacks (see Definition 6.9) and moreover we will assume that predicate ϕ is decidable. There are basically three ways how to solve vulnerability to timing attacks except putting a system off-line. First, redesign the system, put some monitor or supervisor which prevents dangerous behavior which could leak some classified information (see, for example, [14]) or hide this leakage by inserting some time delays between system's action (see [22, 33] for general insertion functions for non-deterministic finite automata). Now we will define and study this last possibility. First, we need some notation. For $w, w' \in Actt^*$ and $w = x_1.x_2 \ldots x_n$ we will write $w \ll_S w'$ for $S \subset Actt$ iff $w' = x_1.i_1.x_2 \ldots x_n.i_n$ where $i_k \in S^*$ for every k, $1 \leq k \leq n$. In general, an insertion function inserts additional actions between original process's actions (given by trace w) such that for the resulting trace w' we have $w \ll_S w'$ and w' is still a possible trace of the process. In our case we would consider insertion functions $\mathcal{F}$ (called time insertion functions) which insert only time actions i.e. such functions that $w \ll_{\{t\}} \mathcal{F}(w)$. Results of an insertion function depends on previous process behaviour. We can define this dependency similarly as it is defined for observation functions.

Definition 6.10 (*Time Insertion function*) Any function $\mathcal{F} : Actt^\star \to Actt^\star$ is an insertion function iff for every $w \in Actt^*$ we have $w \ll_{\{t\}} \mathcal{F}(w)$. It is called static/dynamic/orwellian/m-orwellian ($m \geq 1$) if the following conditions hold respectively (below we assume $w = x_1 \ldots x_n$):

- static if there is a mapping $f : Actt \to \{t\}^*$ such that for every $w \in Actt^\star$ it holds $\mathcal{F}(w) = x_1 . f(x_1) . x_2 . f(x_2) \ldots x_n . f(x_n)$,
- dynamic if there is a mapping $f : Actt^\star \to \{t\}^*$ such that for every $w \in Act^\star$ it holds $\mathcal{F}(w) = x_1 . f(x_1) . x_2 . f(x_1 . x_2) \ldots x_n . f(x_1 . \ldots . x_n)$,
- orwellian if there is a mapping $f' : Actt \times Actt^\star \to \{t\}^*$ such that for every $w \in Actt^\star$ it holds $\mathcal{F}(w) = x_1 . f(x_1, w) . x_2 . f(x_2, w) \ldots x_n . f(x_n, w)$,
- m-orwellian if there is a mapping $f' : Actt \times Actt^\star \to \{t\}^*$ such that for every $w \in Actt^\star$ it holds $\mathcal{F}(w) = x_1 . f(x_1, w_1) . x_2 . f(x_2, w_2) \ldots x_n . f(x_n, w_n)$, $w_i = x_{max\{1,i-m+1\}} . x_{max\{1,i-m+1\}+1} \cdots x_{min\{n,i+m-1\}}$.

Note that contrary to general insertion function (see [22, 33]) inserting time actions is much simpler due to transition rules $A1, A2, S$. The purpose of time insertion function is to guaranty security with respect to process opacity. Let P is prone to S-timing attack or L-timing attack or L- with respect to $\mathcal{O}$ and ϕ. If $\mathcal{O}$ and ϕ is clear from a context we will omit it. Now we define what it means that process can be immunized by a time insertion function.

First we will define immunizability with respect to S-timing attacks and than with respect to L-timing attacks.

Definition 6.11 We say that process P can be L-immunized with respect to the predicate ϕ over $Actt^\star$ and the observation function $\mathcal{O}$ if for every sequence w, $w \in Tr(P)$ such that $\phi(w)$ holds and $\mathcal{O}(w) \neq \epsilon$, and there does not exist a sequence w', $w' \in Tr(P)$ such that $\neg \phi(w')$ holds and $\mathcal{O}(w) = \mathcal{O}(w')$ there exists w_t, $w_t \in Tr(P)$ such that $w \ll_{\{t\}} w_t$ and $\phi(w_t)$ either does not hold or holds and in this case $\mathcal{O}(w_t) \neq \epsilon$, and there exists a sequence w'_t, $w'_t \in Tr(P)$ such that $\neg \phi(w'_t)$ holds and $\mathcal{O}(w_t) = \mathcal{O}(w'_t)$.

If $\mathcal{O}$ and ϕ are clear from the context we say just that process P can be L-immunized.

Definition 6.12 We say that process P can be S-immunized with respect to a predicate ϕ over $Actt^\star$ and the observation function $\mathcal{O}$ if for every P', $P \xrightarrow{w} P'$ such that $\phi(P')$ holds and there does not exists P'' such that $P \xrightarrow{w'} P''$ for some w' such that $\mathcal{O}(w) = \mathcal{O}(w')$ and $\phi(P'')$ does not hold, there exists w_t, $w \ll_{\{t\}} w_t$ such that $P \xrightarrow{w_t} P''$ and there exists P''' and w'', such that $P \xrightarrow{w''} P'''$ such that $\neg \phi(P''')$ holds and $\mathcal{O}(w_t) = \mathcal{O}(w'')$.

In Fig. 6.1 process S-immunization is depicted.

Now we define observation functions which are not sensitive to τ action, i.e. an occurrence of τ in sequence w has no influence how w is observed.

Fig. 6.1 Process
immunization

$$P \xRightarrow{w} \phi(P') \qquad \mathcal{O}(w)$$
$$\|$$
$$P \xRightarrow{w'} \neg\phi(P'') \qquad \mathcal{O}(w')$$
$$P \xRightarrow{w_t} \phi(P') \qquad \mathcal{O}(w_t)$$
$$\|$$
$$P \xRightarrow{w''} \neg\phi(P''') \qquad \mathcal{O}(w'')$$

Definition 6.13 We say that observational function $\mathcal{O}$ is not sensitive to τ action iff $\mathcal{O}(w) = \mathcal{O}(w|_{At})$ for every $w \in Act^*$. Otherwise we say that $\mathcal{O}$ is sensitive to τ action.

The following notes illustrate processes which can and cannot be L-immunized or S-immunized.

Note 6.3 Process P, $P = t.R + (a.Q|\bar{a}.Nil) \setminus \{a\}$, cannot be S-immunized if $\mathcal{O}$ is sensitive to τ action and $\phi(R)$, $\neg\phi(Q)$ hold. An immunization should put a time delay into the trace performed by the right part of the process P but this subprocess cannot perform t action due to the inference rule Pa before communication by means of channel a.

Note 6.4 Process P, $P = h.\tau.t.\tau.Nil + (a.b.Nil|\bar{a}.\bar{b}.Nil) \setminus \{a, b\}$, cannot be L-immunized if $\mathcal{O}$ is sensitive to τ action and $\phi(w)$ is such that it is true iff w contains action h. An immunization should put a time delay into the trace performed by the right part of the process P but this subprocess cannot perform t action due to the inference rule Pa.

Note 6.5 Let $P = l.h.t.l.Nil + l.l.Nil$ and $\mathcal{O}(l) = l, \mathcal{O}(h) = \epsilon, \mathcal{O}(t) = t$ i.e. an observer does not see only action h. Let $\phi(w)$ holds if it contains action h. Then $P \notin Op_{\mathcal{O}}^{\phi}$ since after performing trace $l.h.t.l$ for which ϕ holds and which is observed as $l.t.l$ there is no other trace of P observed as $l.t.l$ for which ϕ does not hold. But if we apply 2-orwellian time insertion function which puts between two subsequent actions l time action t then the resulting process belongs to $Op_{\mathcal{O}}^{\phi}$.

Note 6.6 Let $P = l.t.l.b.Nil + l.l.Nil$ and $\mathcal{O}(l) = l, \mathcal{O}(b) = b, \mathcal{O}(t) = t$ i.e. an observer see every action. Let $\phi(Q)$ holds if Q can perform action b. Then $P \notin POp_{\mathcal{O}}^{\phi}$ since after performing trace $l.t.l$ it reaches the state for which ϕ holds there is no other trace of P observed as $l.t.l$ after which f ϕ does not hold. But if we apply 2-orwellian time insertion function which puts between two subsequent actions l time action t then the resulting process belongs to $POp_{\mathcal{O}}^{\phi}$.

In some special cases, processes can be L-immunized and also S-immunized. as it is stated by the following proposition.

Proposition 6.1 *Let P is prone to L-timing attack or S-timing attack with respect to $\mathcal{O}$ and ϕ. Let $\tau \notin L(P)$ and $\mathcal{O}$ is static. Then P can be L-immunized and S-immunized.*

Proof From transition rule Pa we know that we can always insert t actions since $\tau \notin L(P)$ and since $\mathcal{O}$ is static any process can be L-immunized and S-immunized.

$\square$

Another problem, which causes that processes cannot be immunized, is related to the observation of time, namely, if this observation is context-sensitive, as it is stated in the following note.

Case 6.1 Process P, $P = h.R.Nil + t.Q.Nil$ cannot be immunized for dynamic $\mathcal{O}$ such that $\mathcal{O}(w.t^*.h.t^*.w') = \mathcal{O}(w.w')$, $\mathcal{O}(w.l.w') = \mathcal{O}(w).l.\mathcal{O}(w')$, $\mathcal{O}(w.t.w') = \mathcal{O}(w).t.\mathcal{O}(w')$, and $\phi(R)$, $\neg\phi(Q)$ hold.

Case 6.2 Process P, $P = h.l.Nil + t.l.Nil$ cannot be immunized for dynamic $\mathcal{O}$ such that $\mathcal{O}(w.t^*.h.t^*.w') = \mathcal{O}(w.w')$, $\mathcal{O}(w.l.w') = \mathcal{O}(w).l.\mathcal{O}(w')$, $\mathcal{O}(w.t.w') = \mathcal{O}(w).t.\mathcal{O}(w')$, and $\phi(w)$ is such that it is true whenever w contains action h.

Now we define L-non-time contextuality formally.

Definition 6.14 Let $w = t^{i_1}.s_1 \ldots s_n.t^{i_{n+1}}$ where $i_j \geq 0$ for $0 \leq j \leq n + 1$ and $s_i \in Act^*$ for for $0 \leq i \leq n$. We say that observational function $\mathcal{O}$ is time L-non-contextual if $\mathcal{O}(w) = t^{i_1}.p_1 \ldots p_n.t^{i_{n+1}}$ where $p_i \in Act^*$.

An observational function $\mathcal{O}$ which is L-non-contextual see elapsing of time regardless of other actions.

For functions that are L-non-contextual and not sensitive to τ action processes can be immunized as it is stated by the following proposition. For the proof see [13].

Proposition 6.2 *Let process P is prone to L-timing attacks with respect to ϕ and time L-non-contextual observation function $\mathcal{O}$ which does not see τ. Then P can be immunized with respect to L-timing attacks.*

Corollary 2 Let $\mathcal{O}$ be a static observation function such that $\mathcal{O}(\tau) = \epsilon$ and $\mathcal{O}(t) = t$. Then process P which is prone to timing attacks with respect to ϕ and observation function $\mathcal{O}$ can be immunized for opacity with respect to timing attacks.

To formulate similar property for S-timing attacks we need a slightly different concept of non-contextuality.

Definition 6.15 We say that observational function $\mathcal{O}$ is time S-non-contextual if $\mathcal{O}_t(w) = \mathcal{O}_t(w')$ for every w, w' such that $w \ll_{\{t\}} w'$.

Proposition 6.3 *Let process P is prone to timing attacks with respect to ϕ and time S-non-contextual observation function $\mathcal{O}$ which does not see τ. Then P can be immunized for opacity with respect to timing attacks.*

Proof The main idea. Let $P \notin POp_{\mathcal{O}}^{\phi}$ but $P \in POp_{\mathcal{O}_t}^{\phi}$. This means that there exists a sequence w and P' such that $P \xrightarrow{w} P'$, $\phi(P')$ holds and there does not exist P'' such that $P \xrightarrow{w'} P''$ for some w' such that $\mathcal{O}(w) = \mathcal{O}(w')$ and $\phi(P'')$ does not hold.

Suppose that P cannot be immunized, i.e. for every w_t, $w \ll_{\{t\}} w_t$ if $P \xrightarrow{w_t} P'''$, $\phi(P''')$ holds, there does not exist P''''' such that $P \xrightarrow{w''} P''''$ for some w'' such that $\mathcal{O}(w_t) = \mathcal{O}(w''')$. But due to our assumption we have $\mathcal{O}_t(w_t) = \mathcal{O}_t(w)$ and hence it is with contradiction that P is prone to timing attacks.   □

Corollary 3 Let $\mathcal{O}$ be a static observation function such that $\mathcal{O}(\tau) = \epsilon$ and $\mathcal{O}(t) = t$. Then process P which is prone to S-timing attacks with respect to ϕ and observation function $\mathcal{O}$ can be immunized for process opacity with respect to timing attacks.

Till now we have investigated the properties of observational function and how they influence timing attacks. Now we will concentrate on the properties of predicate ϕ over processes. First, we define what it means that a predicate is time-sensitive.

Definition 6.16 We say that predicate ϕ is time sensitive iff whenever $\phi(P)$ holds for P then there exists $n, n > 0$ such that $P \xrightarrow{t^n} P'$ and $\phi(P')$ does not hold.

Proposition 6.4 *Let process P be prone to timing attacks with respect ϕ and time S-non-contextual observation function $\mathcal{O}$ which does not see τ. Let $\neg\phi$ is not time-sensitive predicate then P can be immunized for S-timing attacks.*

Proof Let $P \xrightarrow{s} P'$ and $\phi(P')$ holds. Since ϕ is time sensitive there exists $n, n > 0$ such that $P' \xrightarrow{t^n} P''$ and $\phi(P'')$ does not hold. Hence process P can be immunized for S-timing attacks. by means of an insertion function which inserts n actions t after sequence s.   □

In general, we cannot decide whether the process can be immunized as it is stated by the following proposition. Fortunately, it is decidable for the most important static and m-orwellian observation functions.

Proposition 6.5 *Immunizability is undecidable i.e. it cannot be decided whether P can be immunized.*

Proof We show a proof of S-immunizability. Let T_i represents i-th Turing machine under some numeration of all Turing machines. We start with generalized process from Note 6.1. Let $P = h.l.R + \sum_{i \in N} t^i.l.Q$. Let $\mathcal{O}(w.h.t^i.w') = \mathcal{O}(w.w')$ whenever T_i halts with the empty tape and $\mathcal{O}(w.h.t^i.w') = \mathcal{O}(w).t^i.\mathcal{O}(w')$ otherwise. It is easy to check that the S-immunization of P is undecidable. By similar the technique we can prove the same for L-immunizability.   □

We will now examine the conditions under which immunizability could be decided. We will model an insertion function by special process F (called time insertion process) which will be put between process P and its environment, see Fig. 6.2. To do so, first, we rename all original actions from Act to actions indexed by symbol a, i.e. $f : Act \to Act_a$ such that $f(x) = x_a$. We require that for every $w \in Tr(P)$ there exists w' such that $w' \in Tr((P[f]|F) \setminus Act_c)$ and $w \ll_{\{t\}} w'$.

To compute m-orwellian insertion function we need only finite memory and hence we have the following Lemma.

Fig. 6.2 Insertion function

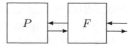

Lemma 6.1 *For every m-orwellian time insertion function there exists a correspond-ing finite-state time insertion process.*

The first condition we put on the insertion process is such that every output of the system with insertion function applied to process P could be also an outcome of P itself.

Definition 6.17 We say that time insertion process is admissible iff for every P and for every w, if $w \in Tr((P[f]\|F) \setminus Act_a)$ then $w \in Tr(P)$.

Definition 6.18 We say that process P is L-immunized (S-immunized) by insertion process F with respect ϕ and $\mathcal{O}$ iff $P \notin Op_{\mathcal{O}}^{\phi} (P \notin POp_{\mathcal{O}}^{\phi})$ but $(P[f]\|F) \setminus Act_a \in Op_{\mathcal{O}}^{\phi} ((P[f]\|F) \setminus Act_a \in POp_{\mathcal{O}}^{\phi})$, respectively.

From now on we assume that the validity of a predicate ϕ can be verified by a finite state process i.e. process which performs some special action after perform-ing sequence w for which ϕ holds (see for details [12]). Now we formulate three properties of immunization.

Proposition 6.6 *Let process P could be L-immunized (S-immunized) by m-orwellian time insertion function, ϕ and observation function $\mathcal{O}$. Then there exists finite state admissible insertion process F such that $(P[f]\|F) \setminus Act_a \in Op_{\mathcal{O}}^{\phi} ((P[f]\|F) \setminus Act_a \in POp_{\mathcal{O}}^{\phi})$.*

Proof According to Lemma 6.1 we know that m-orwellian time insertion function can be modeled by finite state process the rest follows from Definitions 6.17, 6.11 and 6.18.   □

Proposition 6.7 *Let process P could be L-immunized (S-immunized) with respect to timing attacks, ϕ and m-orwellian observation function $\mathcal{O}$. Then there exists finite state admissible insertion process F such that $(P[f]\|F) \setminus Act_a \in Op_{\mathcal{O}}^{\phi} ((P[f]\|F) \setminus Act_a \in POp_{\mathcal{O}}^{\phi})$.*

Proof According to the previous proposition we have to show that P could be immu-nized by m-orwellian time insertion function but this follows from the fact that observation function is m-orwellian.   □

Proposition 6.8 *Given finite state process P and finite state insertion process F. It is decidable whether $(P[f]\|F) \setminus Act_a \in Op_{\mathcal{O}}^{\phi} ((P[f]\|F) \setminus Act_a \in Op_{\mathcal{O}}^{\phi})$ for m-orwellian observation function $\mathcal{O}$.*

Proof The proof follows for decidability of opacity and process opacity for finite state processes in the case decidability of corresponding predicate (see [10]) and from the fact that time insertion function is modeled by finite state process.   □

From the previous propositions, we have the following result.

Proposition 6.9 *Immunizability is decidable for static and m-orwellian observation function $\mathcal{O}$.*

Proof According to Proposition 6.2 it is enough to show that observation function $\mathcal{O}$ is time non-contextual observation function and it does not see τ. Clearly, both these properties are decidable for static and m-orwellian observation functions.   □

If a process cannot be (completely) immunized we can employ supervisory control which eliminates such process behavior that could leak some private information on the validity of ϕ. Namely, it restricts those traces which could lead to L-timing or S-timing attacks.

6.4   Supervisory Control and Timing Attacks

Suppose that a process is prone to L-timing attack or S-timing attack and cannot be L-immunized and S-immunized, respectively. To guarantee its safety, we recommend using a supervisor controller which prohibits such process runs (i.e. traces) which could lead to leakage of private information. From now on we assume process which are prone to timing attacks and cannot be immunized. For simplicity, we assume only static observation functions and such that $\mathcal{O}(\tau) = \epsilon$. We now introduce some basic concepts of supervisory control theory. See [27] for more details . Let us assume deterministic finite automaton (DFA) $G = (X, E, \delta, x_0)$, where X is the finite set of states, E is the set of events, $\delta : X \times E \to X$ is the (partial) transition function, $x_0 \in X$ is the initial state. The transition function can be naturally extended to strings of events. The generated language of $G = (X, E, \delta, x_0)$ is defined as $L(G) = \{s, s \in E^*$ such that $\delta(x_0, s)$ is defined$\}$.

The goal of supervisory control is to design a control agent (called a supervisor) that restricts the behavior of the system within a specification language $K \subseteq L(G)$. The supervisor observes a set of observable events $E_S \subseteq E$ and is able to control a set of controllable events $E_C \subseteq E$. The supervisor enables or disables controllable events. When an event is enabled (resp., disabled) by the supervisor, all transitions labeled by the event are allowed to occur (resp., prevented from occurring). After the supervisor observes a string generated by the system it tells the system the set of events that are enabled next to ensure that the system will not violate the specification.

A supervisor can be represented by $Sup = (Y, E_S, \delta_s, y_0, \Psi)$, where (Y, E_S, δ_s, y_0) is an automaton and $\Psi : Y \to \{E' \subseteq E | E_{UC} \subseteq E'\}$ where $E_{UC} = E \setminus E_C$ specifies the set of events enabled by the supervisor in each state.

System G under the control of a suitable supervisor Sup is denoted as Sup/G, and it satisfies $L(Sup/G) \subseteq K$.

Definition 6.19 (*Controllability*) Given a $DFAG$, a set of controllable events E_C, and a language $K \subseteq L(G)$, K is said to be controllable (wrt $L(G)$ and E_C) if

$$\bar{K} E_{UC} \cap L(G) \subset \bar{K}$$

where $\bar{K}$ is the prefix closer of K.

The controllability of K requires that for any prefix s, $s \in K$, if s followed by an uncontrollable event $e \in E_{UC}$ is in $L(G)$, then it must also be a prefix of a string in K.

Definition 6.20 (*Observability*) Given a $DFAG$, a set of controllable events E_C, a set of observable events E_S, and a language $K \subseteq L(G)$, K is said to be observable (wrt $L(G)$, E_S and E_C) if for all $s, s' \in \bar{K}$ and all $e \in E_C$ such that $se \in L(G)$, $s =_S s'$ ($s =_S s'$ means that strings are equal with respect to the set E_S), $s.e \in \bar{K}$.

Observability requires that the supervisor's observation of the system (i.e., the projection of s on E_S) provides sufficient information to decide after the occurrence of a controllable event whether the resultant string is still in $\bar{K}$.

Proposition 6.10 *Let $K \subseteq L(G)$ be a prefix-closed nonempty language, E_C the set of controllable events and E_S the set of observable events. There exists a supervisor Sup such that $L(Sup/G) = K$ if and only if K is controllable and observable.*

Now we will concentrate on enforcing opacity and process opacity, namely, how to guarantee that there is no leakage of information on the validity of ϕ with respect to opacity. Now let us assume that process P is not secure with respect to opacity (case of process opacity could be treated similarly) for a predicate ϕ and observation function $\mathcal{O}$ i.e. $P \notin Op_{\mathcal{O}}^{\phi}$. That means that there exists $w \in Tr(P)$ such that $\phi(w)$ holds but there does not exist w', $w' \in Tr(P)$ such that $\mathcal{O}(w) = \mathcal{O}(w')$ and $\neg\phi(w')$ holds (similarly for process opacity). For security reasons we want to prohibit such computation (i.e. trace w), what will be the role for the supervisory control. Formally, let K, $K \subseteq Tr(P)$, is a set of safe observations, i.e. for every $w \in K$, $\phi(w)$ does not hold or if it holds then there exists w', $w' \in Tr(P)$ such that $\neg\phi(w')$ holds and $\mathcal{O}(w) = \mathcal{O}(w')$. Clearly, if $P \in Op_{\mathcal{O}}^{\phi}$ then $K = Tr(P)$ otherwise $K \subset Tr(P)$. The aim of the control is to design a supervisor Sup which will restrict behavior of the original process P in such a way that for the resulting process Sup/P we have $Tr(Sup/P) \subseteq K$. Note that we do not assume any relations among set of actions visible for an intruder, a set of actions visible for a controller and a set of controllable actions, i.e. sets $E_I(E_I = \mathcal{O}(Actt))$, E_S, E_C, respectively, similarly to [31]. Note

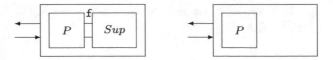

Fig. 6.3 Supervisory Control

that in [6] it is assumed that $E_I \subseteq E_S$ (or $E_S \subseteq E_I$) and $E_C \subseteq E_S$. In [34] $E_I \subseteq E_S$ is assumed and in [32] $E_C \subseteq E_S$ is assumed.

Now we will model supervisory control by means of a special process Sup. Process Sup runs in parallel with P, communicates with environment via actions $Sort(P)$ and internally with P by actions renamed by function f which maps every action a from $Sort(P)$ to a new "ghost" action a' (see Fig. 6.3). The formal definition of a process supervisor is the following.

Definition 6.21 (*Process Supervisor*) Given process P, a process Sup is called supervisor if $L_w(Sup/P) \subseteq L_w(P)$ where $Sup/P = (P[f]\|Sup) \setminus f(Sort(P))$ where $f : Sort(P) \rightarrow Sort(P)'$ where $Sort(P)' = \{x'|x \in Sort(P), x \neq \tau\}$.

We will use a process supervisor to restrict behaviour of the original process in such a way that the resulting process becomes secure with respect to opacity. Similarly it can be done also for process opacity.

Definition 6.22 (*Process Supervisor for Opacity*) Given process P, process Sup is called supervisor for opacity property Op_O^ϕ) (for process opacity POp_O^ϕ)) iff $P \notin Op_O^\phi$ but $Sup/P \in Op_O^\phi$. ($P \notin POp_O^\phi$ but $Sup/P \in POp_O^\phi$.) By $Sup(P, Op_O^\phi)$ ($Sup(P, POp_O^\phi)$) we will denote the set of all supervisors for opacity property (process opacity) Op_O^ϕ (POp_O^ϕ) for a given process P.

Clearly, $Sup(P, Op_O^\phi) \neq \varnothing \neq Sup(P, POp_O^\phi)$ since $Nil \in Sup(P, Op_O^\phi)$ and $Nil \in Sup(P, POp_O^\phi)$. Note that supervisor Nil restricts all behaviour of P which consequently becomes trivially secure. We can formulate some properties of the sets $Sup(P, Op_O^\phi)$ and $Sup(P, POp_O^\phi)$.

To guarantee a minimal restriction of process behavior our aim is to find a maximal process supervisor in the sense that it minimally restricts the behavior of the original process. The formal definition is the following.

Definition 6.23 (*Maximal Supervisor for Process Opacity*) Process $Sup \in Sup(P, Op_O^\phi)$ ($Sup \in Sup(P, POp_O^\phi)$) is called maximal process supervisor for process opacity Op_O^ϕ) (opacity Op_O^ϕ)) iff for every $Sup' \in Sup(P, Op_O^\phi)$ $L(Sup'/P) \subseteq L(Sup/P)$ ($Sup' \in Sup(P, POp_O^\phi)$ $L(Sup'/P) \subseteq L(Sup/P)$).

Clearly, for every process P there exists the maximal supervisor for process opacity Sup, $Sup \in Sup(P, POp_{\mathcal{O}}^{\phi})$ and for opacity Sup, $Sup \in Sup(P, Op_{\mathcal{O}}^{\phi})$.

Unfortunately it is undecidable to verify whether a process Sup is a process supervisors for P and opacity or opacity property as it is stated by the following proposition.

Proposition 6.11 *The property that Sup is a process supervisor for process opacity and opacity for process P is undecidable in general.*

Proof The proof is based on an idea that already process opacity is undecidable (see Proposition 2 in [11]). Suppose that the property is decidable. Let $Sup = \mu X. \sum_{x \in Actt} x'.x.X$ i.e Sup does not restrict anything. We have that Sup is a process supervisor for process opacity for process P iff $P \in Op_{\mathcal{O}}^{\phi}$). Hence we would be able to decide process opacity what contradicts its undecidability. The same holds for opacity. □

Corollary 11 The property that Sup is a maximal process supervisor for process opacity for process P is undecidable in general.

To obtain a decidable variant of the previous property, we insert some restrictions on process or trace predicates. We will do this for process predicates, but trace predicates can be solved similarly.

First, we model predicates by special processes called tests. The tests communicate with processes and produce $\sqrt{}$ action if corresponding predicates hold for the processes. In the subsequent proposition, we show how to exploit this idea for process opacity (for details see [11]).

Definition 6.24 We say that the process T_{ϕ} is the test representing predicate ϕ if $\phi(P)$ holds iff $(P|T_{\phi}) \setminus At \approx_t \sqrt{}.Nil$ where $\sqrt{}$ is a new action indicating a passing of the test. If T_{ϕ} is the finite state process we say that ϕ is the finitely definable predicate.

Suppose that both ϕ and $\neg \phi$ are the finitely definable predicates. Then we can reduce checking whether Sup is a process supervisor for process opacity to checking bisimulation (see Proposition 4. in [11]). Since we can reduce the problem of decidability to finite automata (see [31]) we obtain the following result.

Proposition 6.12 *Let ϕ and $\neg \phi$ are finitely definable predicates and $\mathcal{O}$ is static. The property that Sup is a process supervisor for process opacity for finite state process P and is decidable. Moreover, we can always find a maximal supervisor for process opacity.*

Moreover, for static observational function $\mathcal{O}$ and ϕ and $\neg \phi$ finitely definable predicates there exists finite state maximal process supervisor for process opacity for any finite state process P. This follows from the fact that such observational function can be emulated by finite-state processes since only finite memory is required.

Fig. 6.4 Insertion function
and Supervisor

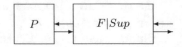

Proposition 6.13 *Let ϕ and $\neg\phi$ are finitely definable predicates and $\mathcal{O}$ is static.*
Then for any finite-state process P there exists finite-state process Sup which is the
maximal supervisor for corresponding process opacity.

To use a supervisor controller to protect systems with respect to S-timing attacks
or L-timing attacks we suggest using it in combination with time insertion functions,
which means to use it only for K, $K \subseteq Tr(P)$ such that for sequences from K there
is no a time insertion function to protect process P with respect to L-timing or S-
timing attacks (see Notes 3 and 4). In fact we can combine an insertion function and
supervisor controller to one process under specific conditions (see Propositions 6.8
and 6.13) as it is depicted in Fig. 6.4.

The supervisor controller is based on sets of observable and controllable events
(actions). Our definition of observation functions is more general because the observability of an action can depend on its context (dynamic or orwellian). However, the
definition of supervisor controller can be suitable extended, which would affect the
definitions and propositions in Sect. 6.4. If an observation (as, for example, m-
orwellian one) can be simulated by a finite-state process , we could obtain similar
decidable properties.

6.5 Conclusions

We have investigated time insertion functions for timed process algebra which
enforce the security with respect to L-timing attacks and S-timing attacks. Time
insertion functions add some delays to the system's behavior to prevent a timing
attack. We study the existence of an insertion function for a given process, a given
observational function, and a predicate over processes or traces. If there is no insertion function that could protect the system against timing attacks we propose to use
a supervisor controller.

In future work, we plan to investigate minimal insertion functions, i.e. such functions which add as little as possible time delays to guarantee process security with
respect to timing attacks. The presented approach allows us to exploit also process
algebras enriched by operators expressing other "parameters" (space, distribution,
networking architecture, power consumption, and so on). Hence we could obtain
security properties that have not only theoretical but also practical value. Moreover,
we can use similar techniques as in [20] to minimize time, as well as other resources,
added to the process's behavior.

Acknowledgements This work was supported by the Slovak Research and Development Agency
under the Contract no. APVV-19-0220 (ORBIS).

References

1. André, E., Lime, D., Marinho, D., Sun, J.: Guaranteeing timed opacity using parametric timed model checking. ACM Trans. Softw. Eng. Methodol. (2021). https://doi.org/10.1145/3502851
2. Bonneau, J., Mironov, I.: Cache-collision timing attacks against AES. In: Goubin, L., Matsui, M. (eds.) Cryptographic Hardware and Embedded Systems - CHES 2006, pp. 201–215. Springer, Berlin (2006)
3. Bryans, J., Koutny, M., Mazare, L., Ryan, P.: Opacity generalised to transition systems. Int. J. Inf. Sec. **7**, 421–435 (2008). https://doi.org/10.1007/11679219_7
4. Bryans, J.W., Koutny, M., Ryan, P.Y.: Modelling opacity using petri nets. Electronic Notes in Theoretical Computer Science **121**, 101–115 (2005). https://doi.org/10.1016/j.entcs.2004.10.010. https://www.sciencedirect.com/science/article/pii/S1571066105000277. Proceedings of the 2nd International Workshop on Security Issues with Petri Nets and Other Computational Models (WISP 2004)
5. Dhem, J.F., Koeune, F., Leroux, P.A., Mestré, P., Quisquater, J.J., Willems, J.L.: A practical implementation of the timing attack. In: Quisquater, J.J., Schneier, B., (eds.), Smart Card Research and Applications, vol. 1820, pp. 167–182. Springer, Berlin (2000). https://doi.org/10.1007/10721064_15
6. Dubreil, J., Darondeau, P., Marchand, H.: Supervisory control for opacity. IEEE Trans. Autom. Control. **55**(5), 1089–1100 (2010). https://doi.org/10.1109/TAC.2010.2042008
7. Focardi, R., Gorrieri, R., Lanotte, R., Maggiolo-Schettini, A., Martinelli, F., Tini, S., Tronci, E.: Formal models of timing attacks on web privacy. Electron. Notes Theor. Comput. Sci. **62**, 229–243 (2001). https://doi.org/10.1016/S1571-0661(04)00329-9
8. Focardi, R., Gorrieri, R., Martinelli, F.: Information flow analysis in a discrete-time process algebra. In: Proceedings 13th IEEE Computer Security Foundations Workshop. CSFW-13, pp. 170–184 (2000). https://doi.org/10.1109/CSFW.2000.856935
9. Gorrieri, R., Martinelli, F.: A simple framework for real-time cryptographic protocol analysis with compositional proof rules. Sci. Comput. Programm. **50**, 23–49 (2004). https://doi.org/10.1016/j.scico.2004.01.001
10. Gruska, D.P.: Process opacity for timed process algebra. In: Voronkov, A., Virbitskaite, I.B., (eds.), Perspectives of System Informatics - 9th International Ershov Informatics Conference, PSI 2014, St. Petersburg, Russia, June 24–27, 2014. Revised Selected Papers, Lecture Notes in Computer Science, vol. 8974, pp. 151–160. Springer (2014). https://doi.org/10.1007/978-3-662-46823-4_13
11. Gruska, D.P.: Dynamics security policies and process opacity for timed process algebras. In: Mazzara, M., Voronkov, A., (eds.), Perspectives of System Informatics - 10th International Andrei Ershov Informatics Conference, PSI 2015, in Memory of Helmut Veith, Kazan and Innopolis, Russia, August 24–27, 2015, Revised Selected Papers, Lecture Notes in Computer Science, vol. 9609, pp. 149–157. Springer (2015). https://doi.org/10.1007/978-3-319-41579-6_12
12. Gruska, D.P.: Security and time insertion. In: Manolopoulos, Y., Papadopoulos, G.A., Stassopoulou, A., Dionysiou, I., Kyriakides, I., Tsapatsoulis, N., (eds.), Proceedings of the 23rd Pan-Hellenic Conference on Informatics, PCI 2019, Nicosia, Cyprus, November 28–30, 2019, pp. 154–157. ACM (2019). https://doi.org/10.1145/3368640.3368668
13. Gruska, D.P.: Time insertion functions. In: Bellatreche, L., Chernishev, G.A., Corral, A., Ouchani, S., Vain, J., (eds.), Advances in Model and Data Engineering in the Digitalization Era - MEDI 2021 International Workshops: DETECT, SIAS, CSMML, BIOC, HEDA, Tallinn, Estonia, June 21–23, 2021, Proceedings, Communications in Computer and Information Science, vol. 1481, pp. 181–188. Springer (2021). https://doi.org/10.1007/978-3-030-87657-9_14

14. Gruska, D.P., Ruiz, M.C.: Opacity-enforcing for process algebras. In: Schlingloff, B., Akili, S., (eds.), Proceedings of the 27th International Workshop on Concurrency, Specification and Programming, Berlin, Germany, September 24–26, 2018, CEUR Workshop Proceedings, vol. 2240. CEUR-WS.org (2018). http://ceur-ws.org/Vol-2240/paper1.pdf

15. Gruska, D.P., Ruiz, M.C.: Process opacity and insertion functions. In: Schlingloff, H., Vogel, T., (eds.), Proceedings of the 29th International Workshop on Concurrency, Specification and Programming (CS&P 2021), Berlin, Germany, September 27–28, 2021, CEUR Workshop Proceedings, vol. 2951, pp. 83–92. CEUR-WS.org (2021). http://ceur-ws.org/Vol-2951/paper7.pdf

16. Handschuh, H., Heys, H.M.: A timing attack on rc5. In: Proceedings of the Selected Areas in Cryptography, SAC '98, pp. 306–318. Springer, Berlin (1998)

17. Hevia, A., Kiwi, M.: Strength of two data encryption standard implementations under timing attacks. ACM Trans. Inf. Syst. Secur. 2(4), 416–437 (1999). https://doi.org/10.1145/330382.330390

18. Jacob, R., Lesage, J.J., Faure, J.M.: Overview of discrete event systems opacity: models, validation, and quantification. Ann. Rev. Control 41, 135–146 (2016). https://doi.org/10.1016/j.arcontrol.2016.04.015. www.sciencedirect.com/science/article/pii/S1367578816300189

19. Ji, Y., Wu, Y.C., Lafortune, S.: Enforcement of opacity by public and private insertion functions. Automatica 93, 369–378 (2018). https://doi.org/10.1016/j.automatica.2018.03.041. www.sciencedirect.com/science/article/pii/S0005109818301286

20. Ji, Y., Yin, X., Lafortune, S.: Enforcing opacity by insertion functions under multiple energy constraints. Automatica 108, 108476 (2019). https://doi.org/10.1016/j.automatica.2019.06.028. www.sciencedirect.com/science/article/pii/S0005109819303243

21. Keroglou, C., Lafortune, S.: Embedded insertion functions for opacity enforcement. IEEE Trans. Autom. Control 66(9), 4184–4191 (2021). https://doi.org/10.1109/TAC.2020.3037891

22. Keroglou, C., Ricker, L., Lafortune, S.: Insertion functions with memory for opacity enforcement. IFAC-PapersOnLine 51(7), 394–399 (2018). https://doi.org/10.1016/j.ifacol.2018.06.331. www.sciencedirect.com/science/article/pii/S240589631830661X. 14th IFAC Workshop on Discrete Event Systems WODES 2018

23. Kocher, P.C.: Timing attacks on implementations of diffie-hellman, rsa, dss, and other systems. In: Koblitz, N. (ed.) Advances in Cryptology – CRYPTO '96, pp. 104–113. Springer, Berlin (1996)

24. Koeune, F., Koeune, F., Quisquater, J.J., jacques Quisquater, J.: A timing attack against Rijndael. Tech. rep., Technical Report CG-1999/1 (1999)

25. Köpf, B., Smith, G.: Vulnerability bounds and leakage resilience of blinded cryptography under timing attacks. In: 23rd IEEE Computer Security Foundations Symposium, pp. 44–56 (2010). https://doi.org/10.1109/CSF.2010.11

26. Milner, R.: Communication and Concurrency. Prentice-Hall Inc, USA (1989)

27. Ramadge, P., Wonham, W.: The control of discrete event systems. Proc. IEEE 77(1), 81–98 (1989). https://doi.org/10.1109/5.21072

28. Rashidinejad, A., Reniers, M., Fabian, M.: Supervisory control synthesis of timed automata using forcible events (2021). https://arxiv.org/abs/2102.09338

29. Rebeiro, C., Mukhopadhyay, D.: A formal analysis of prefetching in profiled cache-timing attacks on block ciphers. J. Cryptol. 34, 21 (2015)

30. Song, D.X., Wagner, D., Tian, X.: Timing analysis of keystrokes and timing attacks on ssh. In: Proceedings of the 10th Conference on USENIX Security Symposium - Volume 10, SSYM'01. USENIX Association, USA (2001)

31. Tong, Y., Li, Z., Seatzu, C., Giua, A.: Current-state opacity enforcement in discrete event systems under incomparable observations. Discret. Event Dyn. Syst. 28(2), 161–182 (2018). https://doi.org/10.1007/s10626-017-0264-7

32. Tong, Y., Ma, Z., Li, Z., Seatzu, C., Giua, A.: Supervisory enforcement of current-state opacity with uncomparable observations. In: 2016 13th International Workshop on Discrete Event Systems (WODES), pp. 313–318 (2016). https://doi.org/10.1109/WODES.2016.7497865

33. Wu, Y.C., Lafortune, S.: Enforcement of opacity properties using insertion functions. In: 2012 IEEE 51st IEEE Conference on Decision and Control (CDC), pp. 6722–6728 (2012). https://doi.org/10.1109/CDC.2012.6426760

34. Yin, X., Lafortune, S.: A new approach for synthesizing opacity-enforcing supervisors for partially-observed discrete-event systems. In: American Control Conference, ACC 2015, Chicago, IL, USA, July 1–3, 2015, pp. 377–383. IEEE (2015). https://doi.org/10.1109/ACC.2015.7170765

Chapter 7
Towards an Anticipatory Mechanism for Complex Decisions in a Bio-Hybrid Beehive

Heinrich Mellmann(iD), Volha Taliaronak(iD), and Verena V. Hafner(iD)

Abstract To a certain extent, humans and many other biological agents are able to anticipate the consequences of their actions and adapt their decisions based on available information on current and future states of their environment. The same principle can be applied to enable decision-making in artificial agents. In order to decide on an action, an agent could envision the consequences for each of the actions and then choose the one promising the best outcome. This anticipatory scheme can enable fast decisions in highly dynamic and complex situations, which has been demonstrated in humanoid robots playing soccer. We extend this principle to the scenario of bio-hybrid beehives augmented with robotic actuators, which allow to influence the foraging locations of the bees. We investigate how a bio-hybrid beehive can make decisions and direct the bees in a way which would benefit the the whole ecosystem enabling sustainable beekeeping. We explore the general principles of anticipation and discuss connections to cognitive science and developmental robotics. We present an implementation of a simulator for the behavior of the augmented beehive and present preliminary results demonstrating the feasibility of the anticipatory approach.

7.1 Introduction

With *Bio-Hybrid Beehive* we refer to a symbiotic system consisting of a beehive augmented with robotic components—sensors, actuators and computational resources. These augmentations provide the bio-hybrid beehive with information regarding the state of the colony and the environment, and enables it to influence foraging locations

H. Mellmann (✉) · V. Taliaronak · V. V. Hafner
Humboldt-Universität zu Berlin, Unter den Linden 6, 10099 Berlin, Germany
e-mail: mellmann@informatik.hu-berlin.de
URL: https://hu.berlin/adapt

V. Taliaronak
e-mail: taliarov@informatik.hu-berlin.de

V. V. Hafner
e-mail: hafner@informatik.hu-berlin.de

B. Schlingloff et al. (eds.), *Concurrency, Specification and Programming, Studies in Computational Intelligence 1091*, https://doi.org/10.1007/978-3-031-26651-5_7

of its bees. From that perspective we can consider the *Bio-Hybrid Beehive* to be a robot that is able to perceive its environment and send its bees to certain locations.

This ability can be used to direct bees to specific areas for foraging with higher yield, and prevent them from foraging in hazardous regions or wildlife preserves protecting wild bees. This could improve the well-being of the honeybees, increase the value generated by them, and reduce the competition with wild bee species. This, in turn, could improve the symbiotic relationship between the honeybees, humans and the ecosystem as a whole. To realize this, the *Bio-Hybrid Beehive* requires an appropriate decision process to decide when and where to send the bees. The foraging behavior of the bees is highly complex and is influenced by a wide variety of factors. Many environmental factors need to be taken into account, including complex topography and actions of other beehives. An action of directing bees to a particular location involves significant and complex uncertainty. This makes inferring a decision a challenging task.

To a certain extent, humans and many other biological agents are able to anticipate the consequences of their actions and adapt their decisions based on available information on current and future states of their environment. This makes anticipation a powerful mechanism enabling complex decision-making and behavior, which should also hold true for artificial agents that interact in complex environments. In an artificial agent, e.g., a robot, anticipation can be realized with the help of an internal simulation. To decide on an action, the internal simulation can be used to envision the outcome of available actions and select the one with the most promising result.

An intuitive approach to making decisions in complex scenarios based on anticipation and internal simulation was presented in [32, 33] in the scenario of humanoid robot soccer—RoboCup. There, humanoid robots play soccer autonomously and have to perceive the environment, make decisions and execute actions in real time in a highly dynamic and complex environment of a soccer game. The overall decision process can be split into three basic steps: *predict* the possible outcomes of the available actions; *evaluate* the outcome according to desired criteria; and *select* the action promising the highest value. In the case of robot soccer—we could simulate where the ball would land after different possible kick actions and choose the one where the ball lands as close to the opponent goal as possible. The central part of this scheme is the prediction step. To predict the results of an action we can use a physical simulator calculating the behavior of the ball after the kick and its interactions with the environment. This simulator is then used as a part of the decision process and is referred to as *internal forward simulation*.

The anticipatory decision scheme provides a direct approach to address the complexity of the *Bio-Hybrid* scenario, similar to the soccer scenario described above. The general intuition remains the same: we envision what would happen when the bees are directed to a certain location and choose the location promising the most desirable outcome. For example, to maximize the yield of honey, we predict how much honey will be collected for each possible location that the bees can be directed to, and then we select the location promising the highest yield.

To study decision-making in the context of bio-hybrid beehives we develop a simulation for the foraging behavior of the bees. This simulation can be used as both an environment to study decision mechanisms, and an internal simulator in an anticipatory decision scheme to predict the outcome of possible actions. Preliminary work on such simulation and decision making in *bio-hybrid* beehives was published in [51]. In this work we extend the simulation, formulate experimental scenarios for studying decision mechanisms, and discuss the generalization of the anticipatory approach to those scenarios.

We begin with a literature review of studies involving anticipation in robots in Sect. 7.2. In Sect. 7.3, we discuss the general principle of anticipation and its role in decision-making in artificial agents. We expand and generalize the framework for a decision mechanism based on anticipation from [32, 33]. In Sect. 7.4, we discuss how the principle of anticipation can be applied to realize decision-making in the scenario of a bio-hybrid system consisting of a beehive augmented with technology, which allows it to influence the foraging locations of its bees and provides access to global knowledge, like safe foraging locations or weather forecasts.

7.2 Literature Review

Often the terms *prediction* and *anticipation* are being used interchangeably. One of the first widely cited characterizations of an anticipatory system was given by Robert Rosen in (Rosen, 1979) [44, p. 537] and later defined more formally in (Rosen, 1985) [45, p. 313]:

> I have come to believe that an understanding of anticipatory systems is crucial, not only for biology, but also for any sphere in which decision making based on planning is involved. These are systems which contain predictive models of themselves and their environment, and employ these models to control their present activities.
>
> (Rosen, 1979) [44, p. 537]

To predict means to envision the future based on the current knowledge of the situation, including possible actions taken by the actors involved. Anticipation, on the other hand, means choosing an action based on the envisioned future. In a certain sense, the mechanism of anticipation can be seen as an *inverse* of the mechanism of prediction.

The concept of anticipation has been extensively studied in different scientific disciplines. The following two collections provide a comprehensive overview—*The Challenge of Anticipation: A Unifying Framework for the Analysis and Design of Artificial Cognitive Systems* [39] and *Anticipatory Behavior in Adaptive Learning Systems: From Psychological Theories to Artificial Cognitive Systems* [40]. A brief overview of anticipatory mechanisms in humans and animals, as well as artificial agents, can be found in [34].

More generally, anticipation appears in a wide variety of control and decision algorithms. For example, an algorithm playing a game of chess might predict several opponent's steps in the future and execute its action in accordance with those predictions. Although the principle of anticipation is very general, we will focus on artificial agents with a physical body, also called *embodied artificial agents*.

Winfield and Hafner [54] consider anticipation in embodied artificial agents through the mechanism of *predictive internal models*, which generate a prediction of a particular state in the future. Depending on the scenario, these models can generate predictions regarding the agent's own body, behavior of other agents (humans or robots), and the environment. Such models can be learned (acquired) by the agent or predefined. Both the fixed and the adaptive model can be used to generate a prediction of a particular state in the future.

A *predictive internal model* is a model of the agent's body, its environment and other agents, which is internal to the agent (part of its cognitive process), and which can be used to make decisions on predictions of the future. In essence, this model needs to capture the laws of reality to a sufficient degree. A predictive internal model can be implicit or explicit. An implicit predictive model can allow reasoning about the future without explicitly computing the state of the future. As an example consider a probabilistic predictive model, which computes the likelihood of occurrence for a given future state. An *internal simulation* is a predictive internal model that generates explicit predictions of the future by simulating real phenomena.

In the following we give a brief overview over studies of anticipatory mechanisms in artificial agents. The studies are roughly divided in two sections: the studies in the Sect. 7.2.1 focus on how the anticipatory mechanism can be acquired (learned) by the agent similar to humans and other animals; the studies in the Sect. 7.2.2 use anticipation as a tool, and investigate how it can be used as a mechanism to realize complex decisions in realistic scenarios.

7.2.1   Adaptive Internal Models

Humans and animals are able to acquire predictive models from experience through exploration in the infant developmental phase, and maintain them over time. In the field of developmental robotics, researchers borrow from theories of human and animal development. They study how a robot can acquire and maintain a model of its own body and the environment by learning from experience that is collected through exploration in a similar way to infants. The learning can involve predictive models as well as corresponding inference mechanisms to derive an action based on predictions. The predictive models can be learned completely or in parts, and can represent the own body, the consequences of own actions, the dynamics of the environment, or actions of the others.

In robotics, the ability to acquire a predictive model of the own body and to anticipate has been demonstrated in several studies and experiments. One of the earliest implementations in real hardware was presented in [7] by Bongard, Zykov and Lipson. There, a four-legged robot resembling a starfish was able to simulate its locomotion and was therefore able to adapt to changes in the morphology of its body, such as removed leg parts.

Demiris and Khadhouri proposed an architecture based on hierarchical networks of inverse and forward models introduced in [11]. This architecture is able of selecting and executing an action; as well as perceiving it when performed by a demonstrator. The idea of inverse and forward models was adopted by Schillaci and colleagues in [47], where they demonstrated that a humanoid robot is able to learn internal models of sensorimotor relationships through an exploratory phase inspired by infants' body babbling. The acquired models can be used for decision-making in tool-use scenarios, as has been demonstrated in [46].

Matsumoto and Tani [31] use predictive coding and active inference for goal-directed planning of grasping trajectories with only partial knowledge. In [20, 35, 38] information theoretic measures are used to learn a self-model of a robot from experience.

Anticipation has also been studied in scenarios involving human-robot interaction, where the robot was required to anticipate human behavior. In [15], the authors implement the ability to read human intentions in a humanoid robot *iCub*. In [12], *iCub* acquires multi-modal models of collaborative action primitives, which enable it to recognize intended action of a human based on human's gaze.

Pico and collages have demonstrated in [41] that forward models can be used to predict the noise a wheeled robot produces by intended motor actions. The models are learned in a non-supervised manner based on random exploration of the noise produced by the motors, also called random motor babbling. A comparison between the predicted and perceived noise allows the robot to infer information about its environment.

In a purely simulated study [28], humanoid agents learn to play soccer, whereby all aspects are learned in a multi-stage process, beginning with skills to strategic decisions. The results of the analysis show, in particular, that the agents acquire an ability to predict the behavior of opponents and teammates. The analysis also shows a positive correlation between higher performance of an agent and its ability to predict future game states.

In general, robots' ability to acquire a predictive model by itself is an extraordinarily complex task. In most scenarios, the learning requires a large amount of data and an extensive exploration phase. To manage the complexity, in most scenarios discussed above, learning is limited to parts of the model relying on classical approaches, for instance, for object detection in images.

7.2.2 Inference of Decisions Based on Anticipation

To realize a decision mechanism based on anticipation, we essentially need a predictive model and an inference mechanism, which uses the predictive model to make decisions. In practical scenarios, we can use external tools to construct a predictive model. For instance, some approaches use a physical simulator calculating the state of the situation based on physical laws. Such internal forward simulation has already been successfully used as an inference method in robotics.

In a minimalist experiment in [6], a small wheeled robot traverses a corridor with moving obstacles. An agent equipped with anticipatory behavior has shown higher success in avoiding collisions in comparison to reactive agents. In [8], the authors investigate the navigation of wheeled robots in a dynamic environment. They use a simulation approach to envision movements of other agents and pedestrians to avoid dynamic obstacles while moving towards a goal.

In [16] a robot equipped with a soft hand explores objects by moving them. The authors use physical simulation of the interaction between the soft hand of the robot and the manipulated object to predict resulting movement of the object. Predictions are used to select actions maximizing the information about the object's properties. In [25] the authors introduce a pancake-baking robot that plans its actions using a full physical simulation of the outcome of possible actions.

There have been several approaches within the RoboCup community to implement decision mechanisms based on anticipation. Mellmann, Schlotter and Blum formulate in [32, 33] an intuitive scheme based on forward simulation, which allows for fast and robust selection of kick-actions in soccer-playing robots. A number of other works focus on a similar task of selecting an optimal kick-action. In [13], a probabilistic approach is used to describe the kick selection problem which is then solved using the Monte Carlo simulation. In [19], the kick is chosen to maximize a proposed heuristic *game situation score*, which reflects the goodness of the situation. In [1], the authors use an instance based representation for the kick actions, and employ Markov decision process as an inference method. In [36], the authors find that projection of the intention of other players can significantly improve the performance of path-planning algorithms.

7.3   Anticipation and Decision-Making in Artificial Agents

In this section, we discuss how anticipation can be realized in embodied artificial agents. For this, we expand and generalize the framework for a decision mechanism based on anticipation from [32] and [33] by Mellmann and Schlotter, which was used to select kick actions in the scenario of humanoid robot soccer—RoboCup. There, humanoid robots play soccer autonomously and have to perceive the environment, make decisions and execute actions in real time in a highly dynamic and complex environment of a soccer game.

Selecting a kick action can present a significant challenge in a complex situation of a soccer game. The outcome of a particular action may depend on a wide variety of environmental factors, such as the robot's position on the field or the location of other players. In addition, the robot's perception of the situation is often uncertain, noisy and incomplete, and the execution of the actions is subject to noise and uncertainty as well. The anticipatory decision-making scheme presented in [32, 33] provides an intuitive and versatile approach to deal with this complexity.

In general, an anticipatory system is a system that makes decisions based on a prediction of the future. This implies the need for two components: a mechanism for predicting the future, and one to infer decisions based on those predictions. From this we can devise a decision scheme consisting of three basic steps: *predict* the possible outcomes of the available actions; *evaluate* the outcome according to desired criteria; and *select* the action promising the highest value. In the case of robot soccer—we could simulate where the ball would land after different possible kick actions and choose the one where the ball lands as close to the opponent goal as possible. The central part of this scheme is the prediction step. To predict the results of an action we can use a physical simulator calculating the behavior of the ball after the kick and its interactions with the environment. This simulator is then used as a part of the decision process and is referred to as *internal forward simulation*.

In the following two sections, Sects. 7.3.1 and 7.3.2, we discuss a modular architecture of the overall cognition and perception process, and practical considerations arising from an anticipatory decision mechanism. In the second half, in the Sect. 7.3.1, we discuss a possible formalization for a general anticipatory decision scheme.

7.3.1 Perception and Decision-Making in Robotics

In this section, we discuss how the cognitive processes and the decision-making processes are commonly realized in complex real world scenarios.

It is challenging to formulate the complete cognitive process of a robot in a real world scenario as a single monolithic process, for instance, as a stochastic process or a neural network. An example of this is the Partially observable Markov decision process (POMDP). POMDP is a general and powerful framework used to formulate the decision process for an agent in a partially observable environment. It works well in robots in isolated real world scenarios, however, is challenging to generalize to the full system where the complete cognition of the robot is formulated as a single POMDP in an end-to-end manner, because of the resulting high computational complexity. In reality, approximations of POMDPs are used to solve parts of the cognitive process, for instance, self-localization with a Particle Filter [9] or object tracking with a Multi-Hypothesis Kalman Filter [23].

Cognition in a complex robot is usually realized as a number of heterogeneous modules and services, which process and integrate sensory information into models of its environment. Such models might perhaps describe the position of the robot in its environment, positions of other objects, people, or robots. These models are updated over time and might be a result of different types of algorithms. These models, at a fixed time t, are also called *state* of the robot at the time t.

Figure 7.1 illustrates an example of a cognitive process implemented in a humanoid robot playing soccer in RoboCup. The process is divided in modules (round nodes) processing information, for instance, detecting the ball in the image and calculating its coordinates or estimating robot's position on the field. Each of the modules can be realized with a different approach, for instance, a deep neural network could be used for object detection, and a particle filter for the estimation of the robot's position on the field. The results, such as coordinates of a soccer ball detected in the image or position of the robot on the field, are stored and communicated between the modules through representations (rectangular boxes) The decision module, denoted

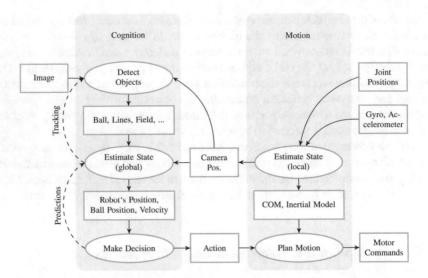

Fig. 7.1 Simplified example of a cognition in a humanoid soccer-playing robot. Round nodes depict modules and services responsible for processing information. Rectangular boxes depict data representations. The overall process is divided in two parts labeled **Cognition** and **Motion**. **Motion** contains critical sensorimotor functionality, e.g., keeping balance, and is executed with higher frequency with guaranteed execution times. **Cognition** groups modules responsible for higher cognitive functionality and is executed with lower frequency

by **Make Decision**, uses the estimated state of the situation to choose and to plan actions.

This subdivision provides high flexibility in choice of algorithms. At the same time, the decision mechanism, depicted by the node **Make Decision**, needs to be able to infer coherent decisions based on heterogeneous data produced by different modules. One of the common approaches to implement the decision mechanism are heuristic rule-based systems, such as widely used CABSL[1] [43] or XABSL[2] [29]. In such cases the goals of the robot are implicitly encoded in manually designed rules, which can lead to high complexity and errors.

The state estimation can be seen as an abstraction layer between a symbolic decision system and uncertain, noisy, ambiguous and incomplete sensory data. The quality of the decision depends on the quality of the state estimation, i.e., how well the estimations correlate with reality, e.g., precision of the estimated position of the robot. One could argue, that in these cases, the complexity and intelligence of the resulting behavior exhibited by the robot stems directly from the state estimation.

A system based on prediction and anticipation, on the other hand, allows to formulate the goals of the robot in an explicit and robust way.

---

[1] https://github.com/bhuman/CABSL (accessed on 8.8.2022).

[2] https://www.sim.informatik.tu-darmstadt.de/xabsl/index.html (accessed on 8.8.2022).

7.3.2   Computational Anticipation

In this subsection, we consider practical aspects of the realization of the anticipatory decision process.

The basic principle of the anticipatory decision process can be summarized in a straightforward way: envision the possible futures resulting from possible actions and choose the action promising the best future.

From a practical standpoint, it is easy to see that an algorithm computing all possible outcomes for all possible actions has exponential growth. To make the process tractable, we need a mechanism to reduce the number of envisioned futures to a few relevant ones. In other words, we need a mechanism to direct the attention to relevant future scenarios.

In order to realize this we need a simulator able to predict future scenarios. When thinking about any kind of simulator (or a predictive model), two main questions arise: concerning the discretization of space and time. For instance, a recurrent neural network predicting a sensory measurement of a pressure sensor runs in fixed time steps, e.g., 10ms and produces numerical values in the working range of the sensor.

The situation becomes less obvious when we consider more complex scenarios on a higher decision level, e.g., a humanoid robot needs to make a decision regarding in which direction to kick the ball. Here, the question concerns the relevant parts of the environment to be simulated and how they to be represented. Evidently, we cannot simulate everything, so a reasonable simplification needs to be done here. For instance, some objects might be described by their position and their movement vectors. We also might limit the simulation only to the objects and agents involved in the situation.

Another important question concerns time. Typical physical simulators execute computations in small equidistant time steps to ensure realistic approximations of physical laws. This, again, might entail large computational effort. Another possibility for discretization in time is based on considering key-events. For instance, when the ball is kicked, the next interesting point in time could be when the ball comes to a standstill or enters the goal. Intermediate movement of the ball can be simplified, as long as influences relevant to the final outcome of the event are captured. This way, time can be split into distinct events and only the state of the situation in those events needs to be simulated.

The final issue that we discuss here is uncertainty. Any prediction in a real-world scenario will have some degree of uncertainty. We can identify several different sources of uncertainty. The first is the uncertainty coming from perception and state estimation, e.g., position of the robot on the field might be shifted, the same goes for the perceived objects. The second component stems from the uncertainty of the robot's actions. For instance, when the robot kicks the ball, the ball doesn't necessarily follow the same trajectory every time, the trajectory instead varies based on the exact location where the ball was hit, orientation of the grass blades, etc. The third source are the discretization and simplification artifacts introduced by the prediction process itself.

7.3.3 Computational Models of Anticipation

In this subsection, we attempt a mathematical formulation for a decision process based on the principle of anticipation. The aim of this subsection is to ground and guide our intuition when talking about anticipation and decision-making. Formulating our ideas in a formal way will illuminate aspects otherwise not easily visible. The aim is not to arrive at a closed theory of anticipation, but rather to structure intuitive ideas and ground them in relation to each other. For our considerations, we will borrow a basic set of tools from function theory, set theory, probability theory and reinforcement learning.

Let's assume that the relevant state at the point in time t can be described by a vector $s_t \in \mathcal{S}$ with the state space $\mathcal{S}$. Let further $a_t \in \mathcal{A}$ describe a possible action at time t in the action space $\mathcal{A}$. In general, we can assume the state and the action to be real vectors with $\mathcal{S} \subseteq \mathbb{R}^n$ and $\mathcal{A} \subseteq \mathbb{R}^m$.

7.3.3.1 Forward Model

An explicit predictive model can be written as a function

$$f : \mathcal{S} \times \mathcal{A} \to \mathcal{S} \tag{7.1}$$

$$(s_t, a_t) \mapsto f(s_t, a_t) := s_{t+1} \tag{7.2}$$

which calculates the envisioned state s_{t+1} based on the current state s_t and the assumed action a_t. The function f can also be called an *explicit forward model*.

More generally, the model can be formulated as an implicit relation φ between the states s_{t+1}, s_t and the action a_t. This relation can be written as

$$\varphi : \mathcal{S} \times \mathcal{A} \times \mathcal{S} \to [0, 1] \tag{7.3}$$

$$(s_t, a_t, s_{t+1}) \mapsto \varphi(s_t, a_t, s_{t+1}) := p_t \tag{7.4}$$

The value of the function φ describes how likely the state s_{t+1} will occur in the future after the execution of the action a_t, given the current state s_t. We could call the function φ an *implicit forward model*.

The main advantage of an explicit model f is that the future state is directly calculated. This can be done, for instance, with forward simulation. Note, f is a deterministic model because it computes a single possible outcome. The implicit model φ can be more expressive. It can reflect the degree to which the future state is likely to occur and, more importantly, several future states can have equally-high likelihood of occurrence. In other words, φ is a probabilistic model. Model φ can be used to compute a likely outcome of an action in a *maximum likelihood* fashion

$$\hat{f}(s_t, a_t) := \underset{s \in \mathcal{S}}{\operatorname{argmax}} \, \varphi(s_t, a_t, s) \tag{7.5}$$

This is, however, not necessarily deterministic, as several possible future states can have the same highest likelihood. In such cases, it will depend on the particular implementation of function argmax, which of the states is selected as the result.

Note, in probability theory, the expected value of s_{t+1} can also be calculated as an average of all possible states weighted with their probabilities. The value calculated this way is also called the *Bayes-hypothesis*. In our context, this approach only makes sense if the state space is continuous and the probability distribution φ is unimodal. Otherwise, in the worst case, it could happen, that the predicted state is very unlikely or might not even be possible, i.e., it might not be in the state space. An example could be a situation where a soccer-playing robot is directly facing a goal post with the ball placed directly in front of it. A forward kick would lead to a collision of the ball with the goal post and result in the ball being deflected to either side of the goal post with equal probability. If we had calculated the expected probabilistic state for this situation, then we would receive a position of the ball directly behind the goal post as an average position of the left and right positions, which is an impossible result.

In a way the forward model captures the geometry and physics of the situation in which the action is executed.

7.3.3.2 Inverse Model

We previously defined forward models, which implicitly or explicitly predict the future state given the knowledge of the current state and an action. The decision problem could be formulated as an inverse of prediction. We want to decide which action to take in order to reach a particular desired state.

Let's consider the explicit forward model f. The current state $s_t \in \mathcal{S}$ is fixed and cannot be changed, thus it could be considered more of a parameter vector, as long as we are considering the fixed time point t and write

$$f_{s_t}(a) := f(s_t, a) \qquad (7.6)$$

Let $s_{t+1} \in \mathcal{S}$ be the desired state. The task of finding an action $a_t \in \mathcal{A}$, that would result in the state s_{t+1} can be formulated as an inverse of f_{s_t}. Please note, that the function f_{s_t} is not necessarily invertible, since there might be several actions that lead to the same target state, i.e., $f_{s_t}(a_1) = f_{s_t}(a_2)$. Thus, the result of the inverse function f^{-1} is a set of all possible action, that would result in x_{t+1}, given the current state s_t.

$$a_t \in f_{s_t}^{-1}(x_{t+1}) := \{a \in \mathcal{A} | f_{s_t}(a) = x_{t+1}\} \qquad (7.7)$$

With this, the decision algorithm computing an action to be executed is not deterministic in general. As a consequence, we need an additional mechanism to select one action. Such a selection can be done based on criteria specific to the action space $\mathcal{A}$. For instance, we could prefer a quicker action, or a less riskier one. In order to express to what degree a given action is desirable, we formulate a risk function

$$r : \mathcal{A} \to \mathbb{R}_+ \tag{7.8}$$

$$a \mapsto r(a) \tag{7.9}$$

The function captures the risks associated with a particular action a. In general, the risk function could be also dependent of the current state as well.

7.3.3.3 Value Function

When making decisions in complex real-world scenarios, there is a set of possible desired states. For instance, in robot soccer it is desired to get the ball inside the opponents goal, the precise place inside the goal box is, however, irrelevant. This subset of goal states $S_G \subseteq S$ might have a compacted shape within the state space S. Additionally, the goal states might be not expressed exactly, but rather the desirability of states might be continuous, i.e., the closer to the opponent goal, the better. This can be captured in a *value function* describing desirability or *value* of every state

$$v : S \to \mathbb{R}_+ \tag{7.10}$$

$$s \mapsto v(s) \tag{7.11}$$

The value function v encodes the value of a particular state with the goal of the systems to maximize the value.

7.3.3.4 Action Selection

Making a decision implies selecting a single action from the set $\mathcal{A}$ to be executed. Previously, we formulated two criteria for the selection of an action, which are captured in the risk and value functions r and v. The aim of the decision algorithm is to minimize r and to maximize v over all possible future states. To formulate this, we can define a *utility function* combining the risk of the action and the value of the state.

$$u : S \times \mathcal{A} \to \mathbb{R}_+ \tag{7.12}$$

$$(s, a) \mapsto u(s, a) := v(s) - r(a) \tag{7.13}$$

With this, the decision mechanism based on the explicit predictive model f can be formulated as a function

$$s : S \to \mathcal{A} \tag{7.14}$$

$$s_t \mapsto a_t := s(s_t) := \underset{a \in \mathcal{A}}{\operatorname{argmax}}\, u(f(s_t, a), a) \tag{7.15}$$

The selection function s assigns each state s_t the action a with the highest value $v(a)$. For s to be well defined we have to assume that the action space $\mathcal{A}$ is not empty. Essentially, function s searches for the action a_t that promises to lead to the future state s_{t+1} with the highest value $v(s_{t+1})$. Function s can be non-deterministic and can have more than one solution, as multiple future states can have the same highest value and different actions might lead to the same future state. In other words, the functions v and f are not necessarily invertible with respect to the argument a.

The state-action space $\mathcal{S} \times \mathcal{A}$ covers all possible combinations of states and actions $(s, a) \in \mathcal{S} \times \mathcal{A}$ and the function f predicts a future state for all combinations. In a real world scenario not all actions must necessary be possible or meaningful in all states. The function f describes what would happen if the action would be attempted. Depending on the concrete scenario the future state might remain the same, meaning $f(s, a) = s$, or result in state with lower value $v(s_{t+1})$ representing the failed attempt. For instance, picture a soccer robot lying face-down on the floor after a fall, in this state executing a kick is not possible and an attempt might be damaging to the robot. If the kick is attempted, then the situation would remain the same, or perhaps contain a malfunctioning knee joint and decreased energy level. In this scenario the kick action would have a significantly lover value compared to a *stand-up-motion* or *do-nothing* action.

Let's consider the combined function $Q := v \circ f$ evaluating state-action pairs

$$Q : \mathcal{S} \times \mathcal{A} \to \mathbb{R}_+ \tag{7.16}$$

$$(s_t, a_t) \mapsto r := (v \circ f)(s_t, a_t) = v(f(s_t, a_t)) \tag{7.17}$$

Using terminology from Reinforcement Learning (RL) we see that the function Q corresponds to the *state-action value function* used in Q-Learning [53]. Function Q predicts the expected value (reward) to be received when the action a_t is executed in the state s_t. The formulation in RL does not usually split function Q in the subcomponents of prediction f and evaluation v. Instead, state-action function Q is formulated in a parameter-free manner, e.g., grid of neural network. This means that the ability to predict state transitions, and to evaluate the states are trained simultaneously. We see here that a basic formulation of the decision problem in RL implies anticipation as a basis for making decisions. One of the most recent studies on application of RL published in [28] shows that the ability to predict future states of the situation emerges in humanoid agents trained to play soccer in simulation. Moreover, this ability positively correlates with agent's performance.

Let's consider the implicit formulation of the decision algorithm.

$$s_i : \mathcal{S} \to \mathcal{A} \tag{7.18}$$

$$s_t \mapsto a_t := s_i(s_t) := \operatorname*{argmax}_{a \in \mathcal{A}} \int_{\mathcal{S}} \varphi(s_t, a, s) \cdot v(s) \mathrm{d}s \tag{7.19}$$

Given action a and current state s_t, the function $\varphi(s_t, a, s)$ describes the likelihood of the future state s. The integral $\int_{\mathcal{S}} \varphi(s_t, a, s) \cdot v(s)$ describes the expected value of the action a. In case $\varphi(s_t, a, s)$ is a probability function this corresponds to the expected value

$$\mathbb{E}(a|s_t) = \int_S \varphi(s_t, a, s) \cdot v(s)\mathrm{d}s \qquad (7.20)$$

7.3.4 Calculating Anticipation

To make a decision in a concrete scenario, we need to compute values for the Fig. 7.19. In general, this can pose a considerable challenge, as the models can take on complicated shapes making explicit calculation intractable.

One possible way to overcome this challenge is approximation through sampling. Sampling is widely used to implement various probabilistic methods in robotics, specifically for state estimation. Particle-Filters used for self-localization are one such examples. For an action $a \in \mathcal{A}$ and a fixed current state s_t we draw a fixed number N of samples for the possible future states with respect to the (density) function $\varphi(s_t, a, \cdot)$. The set of those samples comprises a *hypothesis* for the future state:

$$\mathcal{H}_a^N := \{s_i \in \mathcal{S} | s_i \sim \varphi(s_t, a, \cdot) \wedge 0 \le i \le N\} \qquad (7.21)$$

7.4 Decision-Making in a Bio-Hybrid Beehive

A bio-hybrid beehive is a beehive equipped with additional sensors, actuators and computational resources, which allow the bee colony to better cope with adverse factors in challenging environments and eventually reduce competition for resources between honeybee colonies and other bee species. This augmented beehive can be considered a robot able to make autonomous decisions and act on them.

In this section, we discuss how the principle of anticipation can be applied for decision-making in a scenario involving bio-hybrid beehives. We build and expand on our preliminary work published in [51], where we introduced a heuristic rule-based process for active selection of foraging sources in a bio-hybrid beehive. The approach was implemented and tested using an agent-based simulation for the foraging behavior of the bees.

In the following, we first introduce bio-hybrid beehives. Afterwards, we present an abstract simplified simulation for the behavior of bees, which can be used as an internal simulation in an anticipatory decision-process of a bio-hybrid beehive. In the third part we discuss example scenarios and formulate the decision process. We close the section with remarks on experiments and implementation.

7.4.1 *Intelligent Bio-Hybrid Beehives as Autonomous Robots*

In this section, we provide a brief introduction in specific areas of research of honeybees to provide grounding for our later investigations. Specifically, we illuminate those studies which can be a basis for a connection between the discipline of robotics and studies of bees.

There is a large variety of bee species and some of them have been successfully domesticated by humans. We will refer to the domesticated bees as *honeybees*. Honeybees are primarily used in two major areas: honey production and pollination services in agriculture. In the latter, beehives are intentionally placed in proximity to agricultural fields that require pollination in order to increase yield. Extensive proliferation of honeybees can make them into an invasive species, displacing wild bee species in competition for limited habitable space, as well as foraging resources, such as nectar and pollen. Examples for studies investigating the role of honeybees in extirpation of wild bees can be found in [21] and more recently in [42].

The European Project HIVEOPOLIS [10, 22] aims at developing a true bio-hybrid symbiotic system. The bio-hybrid beehive will be equipped with a wide range of sensors and processing power to monitor and support the health of the colony.

Additionally, the beehive will be equipped with an internal robotic actuator interacting with the bees in the colony through an imitation of a *waggle dance*. In a bee colony, successful foragers can share information about the foraging locations through a specific *waggle dance* [17, 18]. A prototype of a robot imitating the waggle dance, called *RoboBee*, was introduced in [26]. The robot is installed in the beehive and moves a dummy-bee imitating waggle dance, similar to a real bee. It has been shown, that *RoboBee* was able to encourage a significant number of bees to fly to a specific foraging location. The study in [27] investigates the expected effect of dancing robots on the behavior of the bees in a simulated scenario. On the other hand, a waggle dance performed by the bees can be suppressed to prevent them from flying to an undesired location, for instance, because of the contamination with pesticides. Such suppression can be achieved using vibration actuators embedded in beehive as discussed in [50]. A bio-hybrid beehive is connected to a network that allows sharing information with other bio-hybrid units and provides access to a centralized database with extended information regarding possible flowering locations, locations of other beehives in the neighborhood, protected areas reserved for wild bees, weather conditions, and further relevant environmental information.

With this in mind, a bio-hybrid beehive can be seen as a *robot* able to make autonomous decisions and negotiate with other beehives, e.g., send the bees to a particular region or, potentially, prevent them from harvesting in another.

7.4.2 Simulation

At the core of the decision mechanism is the predictive model which allows to predict the outcome of the beehive's actions. This predictive model can be realized as an internal simulation. In this section, we discuss a simplified agent-based simulation for the behavior of bee colonies that can be used to develop and study decision mechanisms for bio-hybrid beehives.

Bees are able to exhibit complex swarm behaviors like decentralized target selection and workload balancing. The behavior of bees has been extensively studied. Specifically, a wide variety of simulations have been proposed, which allow to synthetically investigate behavioral dynamics of the bee swarms. Schmickl and Crailsheim introduce in [48] a multi-agent simulation, which is able to simulate the dynamics of honeybee nectar foraging. The authors implemented experiments reported in [49] and other works of T. Seeley, who investigated natural decision-making mechanisms within a bee swarm. Multi-agent approach was widely employed to study foraging behavior of the bees, including works by Dornhaus and colleges [14], and Beekman and colleges [4]. Well known models BEESCOUT [2] and BEEHAVE [3] are used for better understanding and exploration of the possible realistic scenarios of natural colony dynamics, bees' searching behavior in habitats with different landscape configuration, as well as interactions between bees within a colony. Another example for an agent-based simulator for honeybee colonies was published in [5]. Finally, in [27], authors investigate the effect of robotic actuator imitating bee dance on the bee colony's foraging decisions in a bio-hybrid beehive using mathematical models.

Although the above models were created from the perspective of studying the behavior of bees, they can be used in an internal simulation of an intelligent bio-hybrid beehive to predict the effects of its actions and the actions of other colonies in its surroundings in order to make decisions on the swarm level based on those predictions.

We investigate how a decision mechanism based on anticipation, as discussed in Sect. 7.3, can be realized with the help of such internal simulation. For this, we implement a simplified simulation for the behavior of bee swarms capturing only the essential aspects. For this, we extend on our preliminary work [51], where we presented an agent-based simulation for foraging behavior of bees.

We implement the simulation as a multi-agent system, since this approach has proven successful in simulating the behavior of beehives. The simulation is implemented as a multi-agent system with three basic types of agents: *bee*, *beehive*, and *field*. The environment is modeled as a discrete squared grid. Each cell of the grid contains an agent *field* and can additionally contain other agents, which can be an agent *beehive* and a number of *bee* agents. All *bee* agents in the same cell as a *beehive* are seen as being inside the *beehive*. The same applies for the *field*—all *bee* agents in the same cell are considered to be foraging in that field. All agents know the coordinates of their cell and have access to the list of other agents in the same cell. Each agent has a location on the grid, i.e., coordinates of the cell in which it is located, a

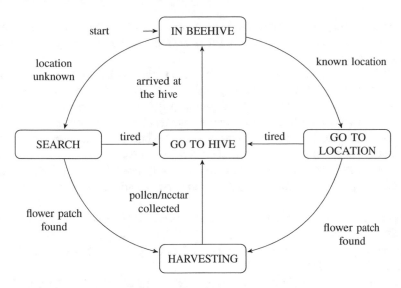

Fig. 7.2 State diagram describing the behavior of a bee agent

number of values describing its internal state, and an activation function, which is called in each step of the simulation. The simulation and the agents are updated in discrete time steps. The activation order of agents is randomized in order to reduce its impact on the model.

The agents *field* and *beehive* represent places with which bees can interact; their own functionality is limited. The agent representing a foraging *field* is implemented in a simple way. The *field* can be in a blooming state, which can change depending on time to simulate blooming periods. The *field* has also a value representing available resources, which can be harvested when the field is blooming. The agent representing *beehive* is mainly responsible for holding information about the state of the colony, like the amount of collected honey. Other than that, the *beehive* does not have its own functionality. It represents a regular beehive without any technical extensions.

The agent *bee* models the foraging behavior of a honeybee. The behavior is implemented as a state-machine shown in Fig. 7.2. We differentiate between five states: IN BEEHIVE, SEARCH, GO TO LOCATION, HARVESTING, and GO TO HIVE. In each state a bee executes a number of actions specific to that state and decides whether to change to another state in the next time step. The bee stays in the current state until conditions for a transition to another state are met. In Fig. 7.2, the self-transitions are not visualized for simplicity. The internal state of a bee is described mainly by the energy that a bee has, a value representing the amount of resources it is carrying, and the current state of the behavior, as described above.

IN BEEHIVE   is the starting state. At the beginning of each simulation all bees are located in their beehives. When inside the beehive, a bee performs three tasks: transferring collected pollen to the hive, recharging energy, and recruiting other bees to its foraging location if foraging was successful. Which bees are recruited and whether recruiting is successful is decided probabilistically, simulating a waggle-dance.

SEARCH   is executed if a bee leaves the hive without a known foraging location. In each step one of the neighboring cells is randomly selected as the next target to move to. The location from the past step is excluded to prevent oscillations between two cells. With each move the amount of energy is reduced.

GO TO LOCATION   works similar to SEARCH, but is more focused on a specific target location. If a bee has a known foraging location, the neighboring cells closer to the target are chosen with higher probability. This means that the bee is gradually pulled towards its target location and can still explore the environment along the route.

HARVESTING   —If the current cell contains a blooming *field*, then transition to HARVESTING is made. In the state HARVESTING the bee collects pollen and nectar if available and leaves after a fixed length of time by transitioning into the state GO TO HIVE.

GO TO HIVE   is implemented in the same way as GO TO LOCATION. In this case the target is the location of the own hive. A bee transitions into GO TO HIVE after HARVESTING, and also from the states SEARCH and GO TO LOCATION in case the bee gets tired and runs low on energy.

Figure 7.3 illustrates an example run of the simulation in a simple scenario with a single beehive and one field with resources. The plot at the bottom in Fig. 7.3 shows the amount of resources (pollen, nectar) collected by bees at each step of the simulation. The collected amount of pollen is represented in an abstract measurement unit. At the beginning of the simulation we see the bees exhibiting the random search behavior. At the time $t = 105$, one of the bees returns with nectar and begins communicating the coordinates of the field to the others. From that time onward, we can observe the behavior of the bees gradually changing to directed harvesting from the field, as the information about its location is propagated among the bees. The change in behavior can also be observed on the shading of the cells, which reflects the number of times a cell was visited by the bees.

To simulate the bio-hybrid beehive, we extend the simulation by adding two new agents *RoboBee* and *BioHybrid*. *RoboBee* is an agent that always stays inside its beehive and participates in recruiting other bees for a certain foraging location. The only task of a *RoboBee* is to hold a provided target location for foraging and to participate the in recruiting of other bees. *BioHybrid* is an extension of the agent *beehive*; it can contain several *RoboBee* agents. Additionally to a *beehive*, *BioHybrid* executes a non-trivial functionality, when activated during the simulation. In each time step, *BioHybrid* has the ability to decide on target foraging locations, which are communicated to its *RoboBee*s. Figure 7.5 illustrates an example for an active selection of foraging locations by a *BioHybrid* equipped with one *RoboBee*. In the left part of Fig. 7.5, no active selection is done; the *RoboBee* is deactivated and the bees behave in the same way as an uncontrolled (regular) beehive. On the right side

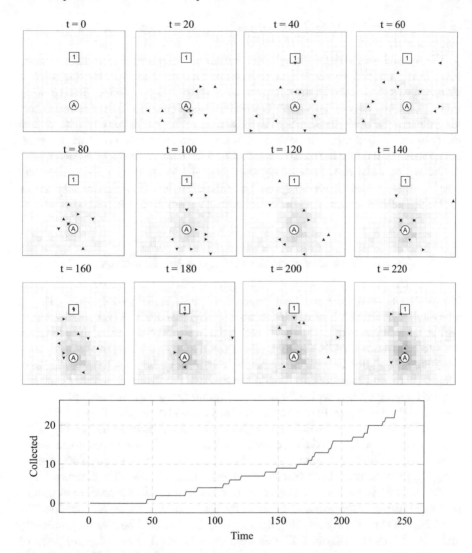

Fig. 7.3 Example for a simulation run with one beehive (circle) and one foraging ground (square). The bees are depicted by the black arrows. The snapshots of the simulated environment are taken every 20 time steps. The shading of the cell indicates how often this cell was visited by a bee. The graph below shows the amount of collected resources (pollen, nectar)

of the Fig. 7.5, the *BioHybrid* actively selected the fixed location of the field further away from the hive.

We focused on the simulation of interactions between beehives as agents, therefore, bees' complex behavior was reduced to foraging behavior. For the sake of simplicity, we do not model the complex social organisation of an individual colony, ignoring the diversity of bees' casts (workers, drones, queen) and food sources (nectar, pollen, water). We implemented two foraging strategies for bees agents: *random foraging search* and *targeted foraging on a known patch of flowers*. In the first case, bees randomly explore the environment around the hive in search of a flower field. In the second case, bees have knowledge about the location of a flower field and head directly towards it for foraging. The simulation has shown promising results for being suitable for studying the decision-mechanism for bio-hybrid beehives.

7.4.3 Anticipation and Decision-Making for Beehives

In this section, we discuss how anticipation can be used to realize decision-making in a bio-hybrid beehive. We want to enable the bio-hybrid beehive to act autonomously within the ecosystem in a way that is beneficial to all involved. The ability to direct bees to certain areas while avoiding others could lead to a wide range of scenarios benefiting all participants of the symbiotic relationship: humans (consumers, beekeepers, farmers), honeybees and wild bee species. For instance, sending bees to a known foraging ground with a high yield of nectar and pollen or plants of a certain kind could lead to higher honey harvest and ensure a specific type of honey. Sending bees to certain locations, which require pollination could increase the quality of the pollination service provided by the bees. Avoiding areas known to be contaminated with pesticides could contribute to bees' health and well-being. Habitats and natural foraging grounds of wild bee species could be protected by avoiding reserved areas.

The bio-hybrid beehive employs a robotic actuator inside the beehive to imitate the waggle dance of the bees and to *recruit* the bees to forage at a certain location. The resulting behavior of the bees is highly complex, since the bees behave autonomously and not all bees necessarily follow the robotic actuator. The actual flight paths of the bees depend highly on the environment and choices of the bees. This means that the actions of a bio-hybrid beehive have complex consequences with a high degree of uncertainty. To make an effective decision in these circumstances, a decision-mechanism will require the knowledge of the environment and understanding of the interactions between the bees the environment and other bee-species. In our previous work [51] we investigated a heuristic rule-based approach, which was shown to work in certain scenarios, but also quickly became difficult to maintain and error-prone due to complexity of the system. To manage the complexity in the behavior of the bees, we investigate a decision-making mechanism based on anticipation as discussed in Sect. 7.3.

We begin with a simple scenario as illustrated in Fig. 7.4. The scenario consists of two beehives—a bio-hybrid beehive (**A**) and a wild beehive (**B**), and two fields (**1**)

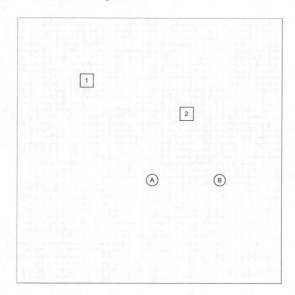

Fig. 7.4 Example scenario: Field 1 is located at the same distance to bio-hybrid beehive A and the wild beehive B. Field 2 is out of reach for wild beehive B, but can be reached by the bio-hybrid A

and **(2)** located in the proximity of the beehives, where bees can forage nectar and pollen. The foraging ground **(2)** is located closer to both beehives, while the other **(1)** is out of reach for the bees from the wild beehive, but still reachable by the bees from the bio-hybrid.

As one of the flower fields is closer to both beehives, it is most likely to be the preferred foraging ground for both colonies. This can lead to a competition for the same resource between the beehives and eventually cause a shortage of food for the wild bee colony. As a result, the wild bee colony might end up unprepared for the next winter period or being forced to migrate to another location. Figure 7.5 (left) illustrates this scenario.

In the second scenario, we assume that the bio-hybrid beehive is aware of the location of the wild beehive as well as other flowering fields in the neighborhood. In order to protect the wild bees, the bio-hybrid beehive could actively encourage its bees to forage at the field **(1)** located at a larger distance. Despite being less attractive, the suggested alternative field still provides enough resources. This approach may decrease the amount of foraged pollen and nectar for the bees of the bio-hybrid beehive because of the distance to the foraging field. Nevertheless, the overall yield of both beehives would still be sufficient and the wild beehive would be protected. In this case, the bio-hybrid beehive collaborates with the wild beehive instead of competing for the same resources. This scenario is illustrated in Fig. 7.5 (right).

In order to make the decision to choose a less convenient option for the sake of common benefit, the bio-hybrid beehive must be aware of the wild beehive's actions. Direct observation of the behavior of the wild bees cannot be realized in a

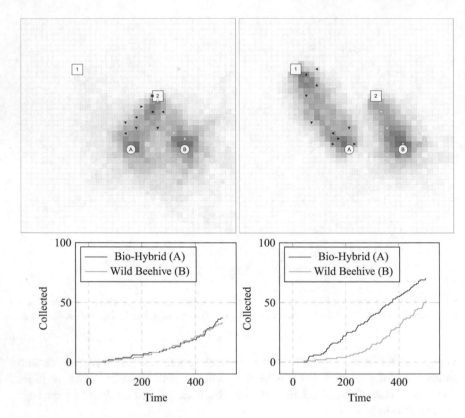

Fig. 7.5 Two scenarios for decision making in bio-hybrid beehives after 500 simulation steps. Bio-hybrid beehive A, a wild beehive B are depicted by circles, and two foraging fields 1 and 2 by squares. Both fields have 100 resources (nectar, pollen). Shading of the cells indicates the number of times the cell was visited by a bee. The number of resources collected by each colony in both scenarios is illustrated by the corresponding plots. Left: with no active control, both beehives A and B choose the closest flower patch 2 and compete for the resources. Right: the bio-hybrid beehive A actively motivates its bees to choose a less convenient foraging location 1 to give the wild beehive space and to improve overall yield

practical way. However, we can use simulation to predict the behavior of the wild bees, assuming we know the location(s) of their beehive(s), distribution of the flowering fields in the neighborhood, as well as a model for their behavior.

We formulate the decision problem more generally. *BioHybrid* could predict the amount of resources collected by each hive for each of the possible foraging locations it could select. Based on that, the target field with the most fair distribution of projected harvest can be selected. As formulated in Sect. 7.3 we need to define components: predictive model, evaluation function, and a selection mechanism.

An *action* executed by the agent *BioHybrid* consists of a target foraging location communicated to the bees in the hive through the *RoboBee*. It is also possible for *BioHybrid* to deactivate *RoboBee*, such that no target foraging location is communicated and the bees are left to forage on their own device. We could call this action to be *neutral*. In this scenario, the set of actions $\mathcal{A}$ consists of all reachable locations in the proximity of the beehive where the bees can be sent.

We consider a time period $[0, T]$ of a fixed length $T \in \mathbb{N}$ and assume that the decision is made only at the beginning of the simulated period, and the action selected by the *BioHybrid* agent remains the same for the whole time. The state s_t is precisely the state of the simulation at the time t. The value function $v(s_t)$ could be formulated as the minimal amount of honey collected by each beehive. That means that the aim is to maximize the least amount of honey collected by a hive, which, in an ideal scenario, would lead to all beehives collecting the same amount of honey.

The simulation presented above can be used as an internal simulator to predict the amount of collected honey for all beehives. Algorithm 5 outlines the three phases of the decision-mechanism: prediction (through simulation), evaluation and selection.

This simple scenario gives an illustration of a situation in which an intelligent bio-hybrid beehive can adapt its behavior to protect a wild beehive, which a regular hive of honeybees would not do. Of course, in reality a bio-hybrid beehive will be confronted with much more complicated scenarios.

Algorithm 5 Action selection based on internal simulation

Data: $S_0, foraging\_fields$
Results: a^*, r^*
$\mathcal{A} \leftarrow foraging\_fields \cup \{none\}$
$consequences \leftarrow \{\}$ $\triangleright$ simulate: run the simulation for each action a
for $a \in \mathcal{A}$ **do**
   $s_0 \leftarrow S_0$
   for $t \in [0, T]$ **do**
      $s_{t+1} \leftarrow simulate(s_t, a)$
   end for
   $consequences \leftarrow consequences \cup \{(a, s_T)\}$
end for
$results \leftarrow \{\}$ $\triangleright$ evaluate: minimal collected ressources by a beehive
for $(a, s_T) \in consequences$ **do**
   $r \leftarrow min(collected\_ressources(s_T))$
   $results \leftarrow results \cup \{(a, r)\}$
end for
$a^* \leftarrow none$ $\triangleright$ select: action with maximal predicted value
$r^* \leftarrow 0$
for $(a, r) \in results$ **do**
   if $r \geq r^*$ **then**
      $a^* \leftarrow a$
      $r^* \leftarrow r$
   end if
end for

7.4.4   Discussion

We implemented two scenarios described in the previous section. In the case of the first scenario, bees from both beehives search for flowering fields through random exploration of their neighbourhoods. As soon as a flowering field has been found, the information about it is communicated to the other bees (in reality it is achieved through waggle dancing). In case of the second scenario, the bio-hybrid beehive has information about food sources in the neighborhood which lie within the maximum flying distance of bees from both beehives. It can estimate at which flowering field competition might eventually happen. If there is such field in the neighbourhood then it searches for the non-competitive fields, estimate which of the non-competitive fields might be the most attractive to the bees, and send them there. In case if all known floral resources are potentially competitive, then bio-hybrid beehive predicts for each of them the maximal number of competitive beehives and suggests one with lowest number of competitors. Of course, the second scenario is an ideal scenario. We should not forget that in reality bees are autonomous agents. They may follow the advertised information and fly to the suggested field but may also ignore it and continue foraging at the same field where the wild bees forage.

The experimental results show that simulation of different strategies can provide data for further analysis in order to define more precise parameters as well as fine tuning of the whole approach.

7.4.5   Remarks on Implementation

Simulations were implemented with the framework Mesa[3] [24]. In comparison to the other well-known simulation tools, like NetLogo[4] [52], Repast[5] [37] and Mason[6] [30], this framework has several competitive advantages. First of all, it is python-based and can be extended with modern python libraries and other python-based tools (e.g., Jupyter Notebook and Pandas tools) in order to create more complex simulations or analyse collected data. The collected data can be stored in a JSON or Pandas DataFrame format for further analysis. Second, Mesa consists of decoupled components, which can be replaced or used independently from each other. Third, visualization is browser-based, which provides additional opportunities for sharing of visualisation via the Internet. Since all components in the Mesa framework are decoupled, visualisation modules can be customized, extended, replaced, or removed.

---

[3] https://mesa.readthedocs.io/ (accessed 08.08.2022).

[4] https://ccl.northwestern.edu/netlogo/ (accessed 08.08.2022).

[5] https://repast.github.io/ (accessed 08.08.2022).

[6] https://cs.gmu.edu/~eclab/projects/mason/ (accessed 08.08.2022).

7.5 Conclusions

Anticipation is a general principle used by humans and other animals to realize complex behaviors. In artificial agents, anticipation can be used to realize a powerful mechanism for inference and decision-making in complex scenarios.

We reviewed, how anticipation is studied in artificial agents and discussed possible formal foundations to ground the intuitive understanding of anticipation in case of decision-making. We discussed the scenario of a bio-hybrid beehive—a beehive equipped with robotic actuators allowing it to interact with bees and to encourage them to forage at certain locations. We implemented a simplified abstract simulation for the behavior of a bio-hybrid beehive. Results of the actions of a bio-hybrid have high complexity due to the high number of involved interactions between the bees and the environment. With the help of the simulation, the complexity can be reduced. Our preliminary experiments show that a decision made by a bio-hybrid beehive can help to protect areas of wild bees. This simulation forms a basis for further studies on a generalized decision mechanism for the bio-hybrid beehive.

To realize the complete decision mechanism for the bio-hybrid beehive we need three components: a predictive model (internal simulation), an evaluation function and a mechanism to select an action based on the value of its predicted outcome.

In its current form the simulator presents a proof-of concept. This is sufficient to study the basic principles of the decision process. In future work we will extend the simulation to better reflect real behavior of the bees and to bring it closer to an application in a real world scenario.

The evaluation function decides which future scenarios are more desirable and thus directly determines the behavior of the agent. For example, the amount of own collected honey, used as a value, would lead to egoistic competitive behavior. The example scenario shown in Fig. 7.5 has demonstrated that it is possible to realize collaborative behavior by considering the envisioned amount of collected honey by others. This opens a question: how can we formulate an evaluation function encouraging collaborative behavior in a more general way? For this we plan to study collaborative and competitive behavior of the beehive in more complex scenarios involving a larger number of bio-hybrid and wild beehives.

Finally, we need to consider the selection mechanism. Because of the high computational complexity, it might not be possible to evaluate all available actions. Possible approaches to solve this could involve sampling.

In conclusion, anticipation in both animals and artificial system is a powerful mechanism for inference and decision-making in complex scenarios and could be studied for a variety of applications.

Acknowledgements This work was supported by the European Union's Horizon 2020 (H2020) research and innovation program under grant agreement No. 824069 (HIVEOPOLIS) and by the Deutsche Forschungsgemeinschaft (DFG, German Research Foundation) under Germany's Excellence Strategy—EXC 2002/1 "Science of Intelligence"—project number 390523135.
The authors would like to thank Vanessa Gorsuch for constructive criticism of the manuscript, which significantly contributed to readability and clarity of this publication.

References

1. Ahmadi, M., Stone, P.: Instance-based action models for fast action planning. In: Visser, U., Ribeiro, F., Ohashi, T., Dellaert, F., (eds.), RoboCup 2007: Robot Soccer World Cup XI, July 9–10, 2007, Atlanta, GA, USA, Lecture Notes in Computer Science, vol. 5001, pp. 1–16. Springer (2007). https://doi.org/10.1007/978-3-540-68847-1_1

2. Becher, M.A., Grimm, V., Knapp, J., Horn, J., Twiston-Davies, G., Osborne, J.L.: Beescout: A model of bee scouting behaviour and a software tool for characterizing nectar/pollen landscapes for beehave. Ecol. Model. **340**(1), 126–133 (2016). https://doi.org/10.1016/j.ecolmodel.2016.09.013

3. Becher, M.A., Thorbek, V.G.P., Horn, J., Kennedy, P.J., Osborne, J.L.: Beehave: a systems model of honeybee colony dynamics and foraging to explore multifactorial causes of colony failure. J. Appl. Ecol. **51**(2), 470–482 (2014). https://doi.org/10.1111/1365-2664.12222

4. Beekman, M., Lew, J.B.: Foraging in honeybees-when does it pay to dance? Behav. Ecol. **19**(2), 255–261 (2007). https://doi.org/10.1093/beheco/arm117

5. Betti, M., LeClair, J., Wahl, L.M., Zamir, M.: Bee++: An object-oriented, agent-based simulator for honey bee colonies. Insects **8**(1), 31 (2017). https://doi.org/10.3390/insects8010031. pubmed.ncbi.nlm.nih.gov/28287445. 28287445[pmid]

6. Blum, C., Winfield, A.F.T., Hafner, V.V.: Simulation-based internal models for safer robots. Front. Robot. AI **4**, 74 (2018). https://doi.org/10.3389/frobt.2017.00074. www.frontiersin.org/article/10.3389/frobt.2017.00074

7. Bongard, J., Zykov, V., Lipson, H.: Resilient machines through continuous self-modeling. Science **314**(5802), 1118–1121 (2006). https://doi.org/10.1126/science.1133687. science.sciencemag.org/content/314/5802/1118

8. Bordallo, A., Previtali, F., Nardelli, N., Ramamoorthy, S.: Counterfactual reasoning about intent for interactive navigation in dynamic environments. In: 2015 IEEE/RSJ International Conference on Intelligent Robots and Systems (IROS), pp. 2943–2950 (2015). https://doi.org/10.1109/IROS.2015.7353783

9. Burchardt, A., Laue, T., Röfer, T.: Optimizing particle filter parameters for self-localization. In: Ruiz-del Solar, J., Chown, E., Plöger, P.G., (eds.) RoboCup 2010: Robot Soccer World Cup XIV, pp. 145–156. Springer, Berlin (2011)

10. CORDIS: Futuristic beehives for a smart metropolis (2019). https://cordis.europa.eu/project/id/824069

11. Demiris, Y., Khadhouri, B.: Hierarchical, attentive, multiple models for execution and recognition (hammer). Robot. Auton. Syst. J. **54** (2005)

12. Dermy, O., Charpillet, F., Ivaldi, S.: Multi-modal intention prediction with probabilistic movement primitives. In: Ficuciello, F., Ruggiero, F., Finzi, A. (eds.) Human Friendly Robotics, pp. 181–196. Springer International Publishing, Cham (2019)

13. Dodds, R., Vallejos, P., Ruiz-del Solar, J.: Probabilistic kick selection in robot soccer. In: Robotics Symposium, 2006. LARS '06. IEEE 3rd Latin American, pp. 137–140 (2006). https://doi.org/10.1109/LARS.2006.334337

14. Dornhaus, A., Klügl, F., Oechslein, C., Puppe, F., Chittka, L.: Benefits of recruitment in honey bees: effects of ecology and colony size in an individual-based model. Behav. Ecol. **17**(3), 336–344 (2006). https://doi.org/10.1093/beheco/arj036

15. Duarte, N.F., Tasevski, J., Coco, M.I., Rakovic, M., Santos-Victor, J.: Action anticipation: Reading the intentions of humans and robots. CoRR (2018). http://arxiv.org/abs/1802.02788

16. Eppner, C., Martín-Martín, R., Brock, O.: Physics-based selection of actions that maximize motion for interactive perception. In: RSS workshop: Revisiting Contact - Turning a problem into a solution (2017). http://www.robotics.tu-berlin.de/fileadmin/fg170/Publikationen_pdf/EppnerMartinMartin17_RSS_WS.pdf

17. von Frisch, K.: Tanzsprache und Orientierung der Bienen. Springer Berlin (1965). https://doi.org/10.1007/978-3-642-94916-6

18. von Frisch, K.: The Dance Language and Orientation of Bees. Harvard University Press, Cambridge, Massachusetts (1967). https://doi.org/10.1007/978-3-642-94916-6. A translation of Tanzsprache und Orientierung der Bienen

19. Guerrero, P., Ruiz-del-Solar, J., Díaz, G.: Probabilistic decision making in robot soccer. In: Visser, U., Ribeiro, F., Ohashi, T., Dellaert, F. (eds.) RoboCup 2007: Robot Soccer World Cup XI, July 9–10, 2007, Atlanta, GA, USA, Lecture Notes in Computer Science, vol. 5001, pp. 29–40. Springer (2007). https://doi.org/10.1007/978-3-540-68847-1_3

20. Hafner, V.V., Kaplan, F.: Interpersonal maps and the body correspondence problem. In: Proceedings of the Third International Symposium on Imitation in Animals and Artifacts, pp. 48–53. University of Hertfordshire, Hatfield (2005)

21. Hudewenz, A., Klein, A.M.: Competition between honey bees and wild bees and the role of nesting resources in a nature reserve. J. Insect Conserv. **17**(1), 1275–1283 (2013). https://doi.org/10.1007/s10841-013-9609-1

22. Ilgün, A., Angelov, K., Stefanec, M., Schönwetter-Fuchs, S., Stokanic, V., Vollmann, J., Hofstadler, D.N., Kärcher, M.H., Mellmann, H., Taliaronak, V., Kviesis, A., Komasilovs, V., Becher, M.A., Szopek, M., Dormagen, D.M., Barmak, R., Bairaktarov, E., Broisin, M., Thenius, R., Mills, R., Nicolis, S.C., Campo, A., Zacepins, A., Petrov, S., Deneubourg, J.L., Mondada, F., Landgraf, T., Hafner, V.V., Schmickl, T.: Bio-Hybrid Systems for Ecosystem Level Effects. In: Proceedings of the ALIFE 2021: The 2021 Conference on Artificial Life, *ALIFE 2021: The 2021 Conference on Artificial Life*, vol. ALIFE 2021: The 2021 Conference on Artificial Life. MIT Press Direct, Virtual (formerly Prague), Czech Republic (2021). https://doi.org/10.1162/isal_a_00396

23. Jochmann, G., Kerner, S., Tasse, S., Urbann, O.: Efficient multi-hypotheses unscented Kalman filtering for robust localization. In: Röfer, T., Mayer, N.M., Savage, J., Saranlı, U. (eds.) RoboCup 2011: Robot Soccer World Cup XV, pp. 222–233. Springer, Berlin (2012)

24. Kazil, J., Masad, D., Crooks, A.: Utilizing python for agent-based modeling: The mesa framework. In: Social, Cultural, and Behavioral Modeling: 13th International Conference, SBP-BRiMS 2020, Washington, DC, USA, October 18-21, 2020, Proceedings, p. 308-317. Springer, Berlin (2020). https://doi.org/10.1007/978-3-030-61255-9_30

25. Kunze, L., Beetz, M.: Envisioning the qualitative effects of robot manipulation actions using simulation-based projections. Artif. Intell. **247**, 352–380 (2017). https://doi.org/10.1016/j.artint.2014.12.004. https://www.sciencedirect.com/science/article/pii/S0004370214001544. Special Issue on AI and Robotics

26. Landgraf, T., Bierbach, D., Kirbach, A., Cusing, R., Oertel, M., Lehmann, K., Greggers, U., Menzel, R., Rojas, R.: Dancing honey bee robot elicits dance following and recruits foragers (2018). https://arxiv.org/abs/1803.07126

27. Lazic, D., Schmickl, T.: Can robots inform a honeybee colony's foraging decision-making? In: Proceedings of the ALIFE 2021: The 2021 Conference on Artificial Life. MIT Press Direct, Virtual (formerly Prague), Czech Republic (2021). https://doi.org/10.1162/isal_a_00397

28. Liu, S., Lever, G., Wang, Z., Merel, J., Eslami, S.M.A., Hennes, D., Czarnecki, W.M., Tassa, Y., Omidshafiei, S., Abdolmaleki, A., Siegel, N.Y., Hasenclever, L., Marris, L., Tunyasuvunakool, S., Song, H.F., Wulfmeier, M., Muller, P., Haarnoja, T., Tracey, B.D., Tuyls, K., Graepel, T., Heess, N.: From motor control to team play in simulated humanoid football. CoRR (2021). https://arxiv.org/abs/2105.12196

29. Loetzsch, M., Risler, M., Jungel, M.: Xabsl - a pragmatic approach to behavior engineering. In: 2006 IEEE/RSJ International Conference on Intelligent Robots and Systems, pp. 5124–5129 (2006). https://doi.org/10.1109/IROS.2006.282605

30. Luke, S., Cioffi-Revilla, C., Panait, L., Sullivan, K.M., Balan, G.: Mason: a multi-agent simulation environment. Simulation **81**(7), 517–527 (2005)

31. Matsumoto, T., Tani, J.: Goal-directed planning for habituated agents by active inference using a variational recurrent neural network. Entropy **22**(5), 564 (2020). https://doi.org/10.3390/e22050564

32. Mellmann, H., Schlotter, S.A.: Advances on simulation based selection of actions for a humanoid soccer-robot. In: Proceedings of the 12th Workshop on Humanoid Soccer Robots,

17th IEEE-RAS International Conference on Humanoid Robots (Humanoids), Madrid, Spain (2017)

33. Mellmann, H., Schlotter, S.A., Blum, C.: Simulation based selection of actions for a humanoid soccer-robot. In: Behnke, S., Sheh, R., Sariel, S., Lee, D.D. (eds.), RoboCup 2016: Robot World Cup XX, pp. 193–205. Springer International Publishing, Cham (2016). http://www.ais.uni-bonn.de/robocup.de/2016/papers/RoboCup_Symposium_2016_Mellmann.pdf

34. Mellmann, H., Schlotter, S.A., Musiolek, L., Hafner, V.V.: Anticipation as a mechanism for complex behavior in artificial agents. In: Bongard, J., Lovato, J., Hebert-Dufrésne, L., Dasari, R., Soros, L., (eds.), Proceedings of the ALIFE 2020: The 2020 Conference on Artificial Life, pp. 157–159. MIT Press Direct (2020). https://www.mitpressjournals.org/doi/abs/10.1162/isal_a_00314

35. Mirza, N.A., Nehaniv, C.L., Dautenhahn, K., te Boekhorst, I.R.J.A.: Anticipating future experience using grounded sensorimotor informational relationships. In: ALIFE2008, pp. 412–419 (2008)

36. Nieuwenhuisen, M., Steffens, R., Behnke, S.: Local multiresolution path planning in soccer games based on projected intentions. In: RoboCup 2011: Robot Soccer World Cup XV, pp. 495–506 (2012)

37. North, M.J., Collier, N.T., Ozik, J., Tatara, E.R., Macal, C.M., Bragen, M., Sydelko, P.: Complex adaptive systems modeling with repast simphony. Complex Adapt. Syst. Model. 1(1), 3 (2013). https://doi.org/10.1186/2194-3206-1-3

38. Olsson, L., Nehaniv, C., Polani, D.: From unknown sensors and actuators to actions grounded in sensorimotor perceptions. Connect. Sci. 18, 121–144 (2006). https://doi.org/10.1080/09540090600768542

39. Pezzulo, G., Butz, M.V., Castelfranchi, C., Falcone, R. (eds.): The Challenge of Anticipation: A Unifying Framework for the Analysis and Design of Artificial Cognitive Systems. Lecture Notes in Artificial Intelligence (LNAI), vol. 5225. Springer, Berlin (2008)

40. Pezzulo, G., Butz, M.V., Sigaud, O., Baldassarre, G. (eds.): Anticipatory Behavior in Adaptive Learning Systems: From Psychological Theories to Artificial Cognitive Systems. Lecture Notes in Artificial Intelligence (LNAI), vol. 5499. Springer, Berlin (2009)

41. Pico, A., Schillaci, G., Hafner, V.V., Lara, B.: How do i sound like? forward models for robot ego-noise prediction. In: 2016 Joint IEEE International Conference on Development and Learning and Epigenetic Robotics (ICDL-EpiRob), pp. 246–251 (2016). https://doi.org/10.1109/DEVLRN.2016.7846826

42. Rasmussen, C., Dupont, Y.L., Madsen, H.B., Bogusch, P., Goulson, D., Herbertsson, L., Maia, K.P., Nielsen, A., Olesen, J.M., Potts, S.G., Roberts, S.P.M., Sydenham, M.A.K., Kryger, P.: Evaluating competition for forage plants between honey bees and wild bees in Denmark. PLoS ONE 16(1), 1–19 (2021). https://doi.org/10.1371/journal.pone.0250056

43. Röfer, T.: Cabsl - c-based agent behavior specification language. In: Akiyama, H., Obst, O., Sammut, C., Tonidandel, F. (eds.) RoboCup 2017: Robot World Cup XXI, pp. 135–142. Springer International Publishing, Cham (2018)

44. Rosen, R.: Anticipatory Systems in Retrospect and Prospect, pp. 537–557. Springer US, Boston (1991 (original 1979)). https://doi.org/10.1007/978-1-4899-0718-9_39

45. Rosen, R.: Anticipatory Systems: Philosophical, Mathematical, and Methodological Foundations. Springer, New York (2012 (orignal 1985)). https://doi.org/10.1007/978-1-4614-1269-4_6

46. Schillaci, G., Hafner, V.V., Lara, B.: Coupled inverse-forward models for action execution leading to tool-use in a humanoid robot. In: Proceedings of the Seventh Annual ACM/IEEE International Conference on Human-Robot Interaction, HRI '12, pp. 231–232. Association for Computing Machinery, New York (2012). https://doi.org/10.1145/2157689.2157770

47. Schillaci, G., Hafner, V.V., Lara, B.: Exploration behaviors, body representations, and simulation processes for the development of cognition in artificial agents. Front. Robot. AI 3, 39 (2016)

48. Schmickl, T., Crailsheim, K.: Costs of environmental fluctuations and benefits of dynamic decentralized foraging decisions in honey bees. In: Anderson, C., Balch, T. (eds.) The 2nd

International Workshop on the Mathematics and algorithms of Social Insects Proceedings, pp. 145–152. Georgia Institute of Technology, Atlanta (2003)

49. Seeley, T., Camazine, S., Sneyd, J.: Collective decision-making in honey bees: How colonies choose among nectar sources. Behav. Ecol. Sociobiol. **28**(4), 277–290 (1991). https://doi.org/10.1007/BF00175101

50. Stefanec, M., Oberreiter, H., Becher, M.A., Haase, G., Schmickl, T.: Effects of sinusoidal vibrations on the motion response of honeybees. Front. Phys. **9** (2021). https://doi.org/10.3389/fphy.2021.670555. https://www.frontiersin.org/article/10.3389/fphy.2021.670555

51. Taliaronak, V., Mellmann, H., Hafner, V.V.: Simulation of interactions between beehives. In: Proceedings of the 29th International Workshop on Concurrency, Specification and Programming (CS&P 2021), pp. 106–112 (2021). http://ceur-ws.org/Vol-2951/

52. Tisue, S., Wilensky, U.: Netlogo: A simple environment for modeling complexity. In: Anderson, C., Balch, T. (eds.) The International Conference on Complex Systems, Boston (2004)

53. Watkins, C.J.C.H., Dayan, P.: Q-learning. Mach. Learn. **8**(3), 279–292 (1992). https://doi.org/10.1007/BF00992698

54. Winfield, A.F.T., Hafner, V.V.: Anticipation in robotics. In: R. Poli (ed.) Handbook of Anticipation: Theoretical and Applied Aspects of the Use of Future in Decision Making, pp. 1–30. Springer International Publishing, Cham (2018). https://doi.org/10.1007/978-3-319-31737-3_73-1

Chapter 8
A Protocol for Reliable Delivery of Streamed Sensor Data over a Low-Bandwidth Wireless Channel

Agnieszka Boruta, Pawel Gburzynski⊙, and Ewa Kuznicka

Abstract We propose a protocol for reliable wireless delivery of a continuous stream of sensor readings to a collection point. Our scheme poses minimalistic demands regarding the sophistication of the RF (Radio Frequency) channel, as well as the amount of memory (buffer space) at the sending device, which can thus be built around a tiny-footprint microcontroller, while practically guaranteeing no losses for as long as the error rate of the channel renders the task formally feasible. The problem arose in the context of a collaborative research project aimed at identification of patterns in IMU (Inertial Mobility Unit) readings collected from working dogs with the intention of applying those patterns to the continuous assessment of the animal's well being. We describe an efficient implementation of our protocol and analyze its performance.

8.1 The Context

This paper deals with a subset of the technical aspects of a wider project, carried out in collaboration between the University of Live Sciences and Vistula University, aiming at researching simple and reliable automated methods for assessing the well-being of working dogs, including service dogs (military, police, disaster-response) as well as assistance dogs (guide, therapy). We were asked to devise a contraption for monitoring the behavior patterns of working canines with the intention of spotting signs of their stress, exhaustion, or any indication of the animal's fatigue or discomfort that would call for the attention of its human companion. The work reported in the

A. Boruta (✉) · E. Kuznicka
Warsaw University of Live Sciences, ul. Nowoursynowska 166, 02-787 Warsaw, Poland
e-mail: agnieszka_boruta@sggw.edu.pl

E. Kuznicka
e-mail: ewa_kuznicka@sggw.edu.pl

P. Gburzynski
Vistula University, ul. Stoklosy 3, 02-787 Warsaw, Poland
e-mail: p.gburzynski@vistula.edu.pl

© The Author(s), under exclusive license to Springer Nature Switzerland AG 2023
B. Schlingloff et al. (eds.), *Concurrency, Specification and Programming, Studies in Computational Intelligence 1091*, https://doi.org/10.1007/978-3-031-26651-5_8

present paper concerns the development of an experimental device for collecting a stream of raw acceleration data to be used for identifying and classifying patterns indicative of the animal's state. More specifically, we describe a transmission protocol for reliable streaming of synchronous data over an unreliable RF channel. The paper is an extended version of our previous report [6]. The extension provides more detail regarding the packaging of data, and includes a performance analysis of the proposed scheme.

The long-term goals of our research probably need no arguments in defense of their compassionate motivation. On top of that, there are also solid remunerative reasons why an effective assessment of the animal's "quality" in providing its service matters. The dog training process is lengthy, complex, and expensive [12, 35], and good quality service/assistance dogs are extremely valuable [27, 32]. This stimulates studies along two lines: (1) to establish reliable criteria for an early assessment of a dog's suitability for a particular kind of work/service [4, 7, 11, 33]; (2) to make sure that the animal is well taken care of and, in particular, any problems related to its work stress and generally health are quickly detected, diagnosed, and addressed [8, 21]. The latter issue can be reformulated as that of an effective communication in the animal-to-human direction [1], as opposed to the more popular direction inherent in dog training.

While anomalies in a dog's behavior indicative of deficiencies in its well-being can be spotted by an expert human (veterinarian, behaviorist) through direct observation, our primary interest is in harnessing to this task sensing devices that, ideally, should be able to detect such problems automatically and signal them to the human supervisor. The issue can be viewed as falling under the more general umbrella of sensor-based diagnostics, e.g., similar to taking and interpreting ECG readings [9]. While it is obviously possible to subject the animal to extensive and authoritative veterinary assessments, including tests and collections of sensor data interpreted by a human expert, we are interested in completely automated monitoring carried out by an inconspicuous and (basically) maintenance-free device being essentially unnoticeable by the animal. Such a device, a wearable *Tag*, would be permanently carried by the dog, e.g., attached to a collar, and would wirelessly convey sensor data to a nearby access point for automated interpretation.

The unobtrusiveness premise of the sensing/monitoring device trades off against the overt information content of the data that it can possibly collect. For example, it may seem worthwhile to try to obtain an EKG/ECG chart of the animal, which is a valuable source of information about the heart activity. One can expect such a chart to reasonably easily translate into a representation of the dog's tiredness or stress. While there exist experimental techniques for collecting this kind of data through "wearable" sensors [14], they overtax our premise by requiring direct access to the animal's skin. Even the less ambitious task of taking reliable heart rate without skin contact proves challenging [15].

Our long-term goal is to investigate how much one can accomplish with a completely unobtrusive device, requiring no skin contact and, preferably, no rigid attachment to the animal (in a specific position or place). The most natural sensor to try in this context is the IMU being a combination of an accelerometer, a gyro, and a compass. Previous studies have reported various degrees of success in using the sensor (most notably the accelerometer component) for diagnosing various behavioral anomalies/problems in dogs [13, 16, 21]. We want to establish criteria for the classification of IMU data into simple signals indicative of some threshold levels of the animal's well-being in relation to its level of fatigue or stress that can be easily communicated to the human companion. The next step will be to built an actual practical device and application based on the outcome of our studies.

The remainder of the paper is organized as follows: Sect. 8.2 outlines the ramification of the project; Sect. 8.3 presents the problem that we aim to solve; Sect. 8.4 describes the data streaming protocol and the relevant aspects of its implementation; Sect. 8.5 discusses the performance of our scheme; finally, Sect. 8.6 sums up our work.

8.2   The Setup

As our project aims at fabricating a practically useful device, it makes sense to start with a view of the target application. Ideally, we should use the same hardware for the experiments and for the final application. As the classification of the animal's activity patterns will be carried out based on the indications of an IMU, the embodiment of the sensor (the weight of the device and the mode of its attachment to the animal) is likely to matter because of its own (inherent) inertia component which will tend to influence the readings. This aspect of the project makes it similar to one of our earlier endeavors [23] where a series of research experiments carried out with an IMU-based sensor provided data to drive the design of a classification algorithm that could be subsequently implanted into the same device to a more practical end.

Having agreed that the experimental device should be identical to the target one, we have to realize that the experimental version of the application (the software run by the device) is going to be drastically different from its target version. The role of the experiments is basically raw data collection. We want to amass a large amount of sensor readings, taken at the maximum rate that we can afford, from various representative animals acting under controllable conditions, where experts can annotate the collected data with authoritative labels indicative of the dog's state. That data will be later used off-line to search for patterns that can guide the classification algorithms to be applied on the data collected by the target incarnation of the device. That part of the research methodology is beyond the scope of the present paper.

The nature of the experiments, demanding that the animals act within environments appearing as close to their natural work conditions as possible, is an additional argument for the experimental devices being identical to the target ones, and it obviously precludes wired connectivity of the devices to external equipment. The footprint of the target device makes it impossible to store large volumes of data directly on it. Besides, the data must be annotated in real time, which makes it natural to use an external computer (a laptop or a tablet) for the actual collection and simultaneous annotation. Therefore, the sensor device is going to *stream* its readings wirelessly

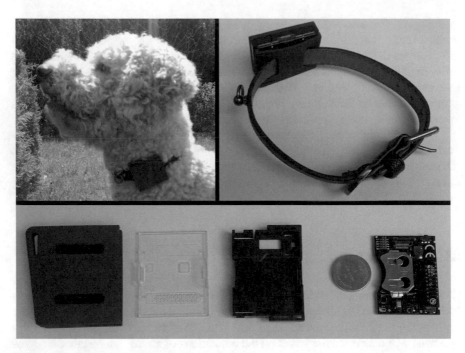

Fig. 8.1 The wearable SensorTag device

to the external computer. The streaming protocol is the aspect of the application
discussed in the remainder of this paper.

The device used for data collection (and envisioned for the target application) is
the CC1350 SensorTag[1] manufactured by Texas Instruments and featuring an ARM-
based CC1350 microcontroller [34]. The SensorTag comes equipped with a number
of sensors including the MPU9250 IMU by TDK InvenSense.[2] The complete device
weights ca. 20 g (including a CR 2032 battery) and its dimensions are $44 \times 32 \times 6$
mm. When attached to a dog's collar, as shown in Fig. 8.1, it is no more obtrusive
than a slightly oversized name tag.

The experimental setup consists of a pair of CC1350-based devices, one of them
being the *Tag* worn by the dog, the other a CC1350 LaunchPad,[3] dubbed the *Peg*,
connected over USB to the computer and acting as the RF access point (data sink)
for the Tag. While the RF module of CC1350 can be configured to operate in Bluetooth mode, thus eliminating the need for a special access point, we opt for the
so-called proprietary mode of the radio which gives us access to the raw channel.
We prefer to circumvent the Bluetooth standard to: (1) increase the range of communication (to provide for a larger separation between the animal and the access

---

[1] https://dev.ti.com/cc1350stk.

[2] https://invensense.tdk.com/products/motion-tracking/9-axis/mpu-9250/.

[3] http://dev.ti.com/launchxl-CC1350.

point), (2) implement our private reliability scheme to increase the delivery fraction of the collected data. The proprietary mode operates within the 915 MHz ISM band offering parameterizable transmission power up to 14 dBm and (raw) transmission rates up to 500 kbps. Data exchanged over the RF channel are organized into packets with the maximum (total) packet length (practically) limited to 60 bytes [18]. The Bluetooth capability of the device will become handy in the target application where the device will be able to communicate directly with a smartphone. The essential classification will then be carried out in the Tag [23], so there will be no need for reliable streaming of large volumes of readings to the access point.

8.3   The Problem

In its most general formulation, independent of the hardware and application context, the problem is that of implementing reliable communication across an unreliable channel. We deal with an asymmetric link where the Tag continuously sends a stream of data to the Peg. Ideally, we would like *all* the data generated by the Tag to (eventually) reach the Peg, in the proper order, such that the complete, timed, and annotated stream of sensor readings produced by the Tag is stored at the collection computer.

Standard solutions to this problem involve a feedback channel operating in the Peg-to-Tag direction, as shown in Fig. 8.2. The simplest of them has been known as the Alternating Bit Protocol [3, 28] and consists in explicitly acknowledging every single packet received by the Peg from the Tag. Various generalizations and improvements upon this simple scheme, including the one underlying TCP, are known under the name of ARQ (Automatic Repeat reQuest) protocols [22, 26]. Their primary objectives are: (1) to reduce the amount of feedback traffic, (2) to improve the continuity of the forward traffic when the losses are low and/or the bandwidth-delay product of the link is large [24].

The problem of *absolutely* reliable data transmission primarily concerns information whose utmost integrity is essential, i.e., the transmission of files. With streaming, the data is assumed to be created continuously and the problem of its reliable delivery receives a different flavor. In many cases the information is not stored at the source in the form of a complete file whose missing fragments could be requested at random by the recipient and retransmitted later. But even if this happens to be the case, the issue of reliable delivery is then conditioned by the real-time character of the communication where the data often become obsolete and useless if not received within a certain time window (like in streaming of voice and/or video). Consequently, while

Fig. 8.2   The channel model

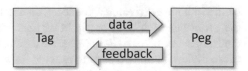

standard streaming applications may insist on high-quality of delivery, they typically are (and usually must be) prepared to deal with innate data loss [2, 31].

Our problem is specific in that: (1) the information collected by the Tag cannot be entirely stored at the source, so the availability of its undelivered fragments is restricted (this aspect makes it similar to standard streaming); (2) there is no issue of real-time playback at the recipient, unlike in streaming audio or video; (3) we can be prepared to deal with some losses, and seemingly have to because of (1), but we would strongly prefer to receive all data. The problem is reminiscent of one we dealt with before [19] (the hardware context was similar), and it may be worthwhile to emphasize the difference. In the case of [19] the entire collected data set was stored at the source Tag, and the problem was formulated as minimizing the time of its *complete* delivery to the Peg. The completeness of delivery was essential, so we were basically interested in reliable and fast file transfers among small-footprint devices. In the present case, we are inclined to deal with occasional data loss (the stream is infinite for all practical purposes), and the issue is to maximize the delivery fraction, which we want to accomplish by trading against the absence of the synchronous delivery requirement characteristic of typical streaming sessions.

If losses are truly unavoidable, one can often simplify the solution by eliminating the feedback channel altogether focusing instead on enhancing the reliability of communication in the physical layer, e.g., through FEC (Forward Error Correction) techniques [29]. The proprietary mode of the RF module of CC1350 comes with a number of options that can be applied to this end. They effectively trade the transmission range and bit rate for reliability and belong to the set of configuration parameters of the RF module that have to be tuned for performance regardless of any other tools. However, depending solely on them would be too restrictive from our point of view. The dynamic nature of the propagation environment, including the variable distance between the Tag and the Peg, would lock the channel into a conservative setting, possibly offering an acceptable (or passable) loss rate for the worst case scenario, while unnecessarily reducing the opportunities for collecting more data in a friendlier environment. Our hope is to achieve a better flexibility with a properly designed feedback scheme where the bandwidth to be sacrificed for reliability is only sacrificed when needed. In this respect, we want our protocol to be adaptive to the environment.

In a high-level discussion of ARQ schemes (and in many implementations of such schemes in the wired world) it is assumed that the feedback channel (Fig. 8.2) is separate from the forward (data) channel. This is to say that the feedback messages do not disrupt the data stream and, in particular, they can be sent at any rate up to some maximum with their impact being solely positive. This is seldom true in wireless communication. Even if the two channels are in fact separate, which they mostly are not, they will tend to interfere. Generally, setting aside a sizable portion of the RF bandwidth for a "frivolous" feedback channel makes that bandwidth unavailable for the *proper* use which is transmitting data. For example, in Bluetooth, e.g., within the framework of an ACL link [5], the essentially single channel must be partitioned (time-divided) into two parts to provide for two-way communication. Realizing that the feedback channel is going to directly coexist with the forward data channel, we would like to make it flexible and, in particular, avoid a rigid pre-allocation of bandwidth. Our goal is to reduce the impact of the feedback channel on data bandwidth to the minimum required by the application and constrained by the fluctuating characteristics of the RF environment.

8.4   The Solution

Owing to the absence of real-time playback requirements, the possibility of unrecoverable losses is solely the consequence of the finite buffers at the Tag. Whatever buffer space is available will be allocated to a shifting window of the collected data. The Peg will be able to request retransmission of those undelivered packets that are still present within the window.

The **Tag side** of the protocol is described by two threads: the generator of blocks of sensor readings (dubbed the generator thread) and the transmitter of those blocks on the RF channel. The blocks are stored in a singly-linked queue, denoted by Q and depicted in Fig. 8.3, whose size is limited. The queue is represented by two pointers: Q_h (the head), and Q_t (the tail). When Q is empty, we have $Q_h = Q_t = null$. One block contains readings to be expedited in a single RF packet. In addition to the readings (whose number is the same for all blocks, see Sect. 8.4.5), a queued block b contains a link to the next block in Q (or $null$ if the block is the tail one) and the sequence number of the block in the stream, which we shall denote by $b{\rightarrow}n$. This number starts with 1 (for the first block generated in a session) and is incremented by 1 for every new block issued by the generator thread, as explained below.

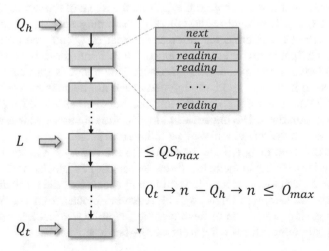

Fig. 8.3 The block queue

The maximum size of the queue (i.e., the maximum number of blocks that it can contain at a time) is determined by the amount of storage available at the Tag and denoted by QS_{max}. The latter can be assumed to be (roughly) equal to M/s where M is the total amount of RAM available at the Tag for storing Q and s is the (fixed) block size. When the streaming operation starts, Q is initialized to empty and N (the next block number to be issued by the generator thread) is set to 1.

8.4.1 The Generator Thread

Every $1/f_s$ s, where f_s is the sampling frequency, a sensor reading is taken. The reading is stored in the *current buffer* denoted by CB. CB is filled by consecutive readings (produced f_s times per second) until it becomes complete (its capacity is reached).

When CB becomes complete, the thread executes a function named add_CB which sets $CB{\rightarrow}n$ to N, increments N by 1, and appends CB at the end of Q (updating Q_t and possibly Q_h as needed). Before adding the new block to the queue the function makes sure that, after the addition of CB, Q will not exceed its two limitations (which are both necessary). **Limitation 1** says that the total number of blocks in the queue is never bigger than QS_{max}. **Limitation 2** requires that $Q_t{\rightarrow}n - Q_h{\rightarrow}n$, i.e., the difference between the first and the last block number in the queue, be less than or equal to O_{max} which we call the *maximum block offset*. The first limitation simply makes sure that Q never exceeds its allotted storage. The second limitation constrains the age difference of the blocks kept in the queue. The primary reason for it is that we want to be able to reference past blocks relative to the current place (block number) within the session, for which we want to restrict the range of the requisite offsets.

Note that CB is appended at the tail of Q becoming new Q_t. To enforce the limitations, add_CB examines the block at the front of Q (pointed to by Q_h) and discards it (in a tight loop) for as long as any of the two limitations would be violated with CB included in the queue. One invariant of the queue is that the numbers of blocks stored in it are strictly increasing (they need not be consecutive as we shall shortly see). Thus, looking at Q_h and the total number of blocks stored in Q is enough to assert the limitations. Having added CB to Q, the function opens a new empty copy of CB which the thread will now be filling from scratch.

The blocks stored in Q will be transmitted to the Peg by the second thread (as explained below), but not immediately discarded, unless new blocks, containing most recent readings, cannot be accommodated into Q because of the limitations. When that happens, the sampling thread will be removing blocks from the front of the queue, thus giving preference to fresh readings. Note that a block is only lost for good when it is removed from the front of Q—because it violates one of the two limitations.

8.4.2   The Transmitter Thread

The thread operates in rounds where it transmits a *train* of blocks to the Peg and then *reverses the channel* for a short while [19] to allow the Peg to acknowledge the train. Once the transmission of a train is started, it is carried out back-to-back with a minimum inter-packet spacing, just to enable the recipient to accept the individual packets from the channel. The Tag is not able to receive anything from the Peg until the complete train has been transmitted. Only then will the Tag listen for the acknowledgment (ACK) packet from the Peg.

A train always consists of the same number of packets (blocks) which we shall denote by T. When commencing a train, the thread starts by scanning consecutive packets from Q (beginning from Q_h) and transmitting them in sequence. When the last packet (the one pointed to by Q_t) has been handled and the train is still incomplete, the transmitter thread will simply wait until the generator thread delivers the next block, i.e., the remaining packets in the train will be sent as new blocks materialize in Q. Note that the blocks are *not* removed from Q as they are being transmitted.

A train packet contains the block number $b{\rightarrow}n$ of the block carried by the packet and the packaged sensor readings copied from the block (Sect. 8.4.5). The block number always allows the Peg to authoritatively place the contents of a received packet within the complete stream of readings, regardless of how many packets have been lost and how the received packets have been misordered with respect to the original stream.

Having completed the current train, the transmitter thread enters a loop in which it expects to receive an ACK packet from the Peg. Within that loop, the thread periodically sends a short EOT (End of Train) packet (to make sure that the Peg has recognized that the train has ended and a response is expected) and waits for a short while for the ACK (which should normally arrive after a minimum delay). The EOT packet carries three items of information: the train number modulo 256 (a single byte) used to match trains to acknowledgments, the number of the last block transmitted in the train (denoted by L), and the back offset (O_b) from L to the oldest packet still held in Q ($O_b = L - Q_h{\rightarrow}n + 1$).

The idea is that having received an EOT packet, the Peg can assess which blocks have been missing (it knows the number of the last block sent by the Tag) and it also knows the minimum number of the block that it can still ask the Tag to retransmit. The reason why the latter is specified as an offset with respect to L is technical: the offset information can be conveyed in two bytes instead of four (needed for a full block number). It illustrates one practical and seemingly mundane aspect of real life in the embedded world where it always makes sense to try to save on individual bytes (or even bits) stored in memory or expedited over the RF channel. Note that the generator thread enforces a limitation reducing the maximum difference between the numbers of blocks stored in Q (Sect. 8.4.1).

The number of the oldest block still available in Q conveyed by the Tag in the EOT packet reflects the state of Q at the moment the EOT packet is constructed and scheduled for transmission. This information may become outdated by the time the Peg builds and transmits its response and, more importantly, by the time that response arrives and is interpreted by the Tag, because Q can be trimmed by the generator thread independently of the transmitter thread. This is OK. The possibility that a block may be irretrievably lost is factored into the scheme. It can happen that the Peg asks for the retransmission of a block that is no longer available. At the end of the next train the Peg will learn (from the new EOT packet) that the block is no more, so it will know that it makes no sense to keep asking for it.

The transmitter thread will continue retransmitting the EOT packet (possibly updating its parameters) until an acknowledgment arrives from the Peg. Normally this should happen right away, but in a pathological scenario, if the connectivity has been broken for a long time, the contents of Q may evolve to the point where L is no longer available. This is why the minimum value of O_b for the situation when Q still contains the block number L is 1. When O_b in an EOT packet is 0, it means that none of the blocks up to (and including) the end of the last train is available any more, so the Peg need not ask for any retransmissions.

8.4.3   The Acknowledgment

The role of the acknowledgment (ACK) packet is to indicate to the Tag which blocks have been missing with respect to the end of the last train and relative to the Peg's knowledge regarding the blocks that it can still hope to receive. Again, mundane technical constraints force us to be frugal about the representation of information within the ACK. For one thing, the entire message should fit into a single packet whose useful (payload) size is limited to 50 bytes. Organizing the ACK message into multiple packets would introduce obscure reassembly problems complicating things beyond practical [19]. Consequently, the ACK format should allow the Peg to maximize the population of (independent) block numbers that it can specify in a single packet to cover worst-case scenarios and, more generally, to minimize the size of a typical ACK packet. As the ACK traffic *interferes* with the otherwise smooth series of back-to-back transmissions of useful data (causing channel reversals [19]), its impact (and the incurred reduction of bandwidth) should be minimized.

One mandatory item carried in an ACK packet is the train number (modulo 256) to which the acknowledgment applies. Any other data included in the packet pertain to the blocks that the transmitter thread of the Tag should *retain* in Q before commencing the next train. In other words, these are the blocks that the Peg wants retransmitted. If the ACK contains no data beyond the single mandatory byte (which is the ideal case), the message to the Tag is simple: erase Q up to and including L, i.e., the end of the last train.

The structure of an ACK packet is best explained by the way the information is interpreted by the Tag upon its arrival. The interpretation is carried out by function *handle_ACK* invoked by the transmitter thread. Following the train number stored as the first byte of the packet, the function interprets consecutive bytes as descriptors of the blocks that have been missing by the Peg and, if still available, should be

Fig. 8.4 Descriptors of missing blocks

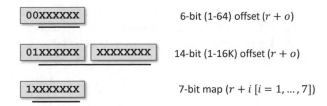

retransmitted. Those blocks are specified within the ACK packet in the increasing order of their numbers.

While interpreting the descriptors, the function stores in r the last block number mentioned by a previous descriptor, to be used as a reference. The value of r is initialized to $L - O_{max} - 1$ where O_{max} is the maximum legal difference between block numbers in Q (see above). The ACK bytes are interpreted in the following way (see Fig. 8.4):

1. If the two most significant bits of the byte are zero, then its remaining six bits are interpreted as a forward offset from r minus 1, i.e., r is incremented by the nonnegative integer value stored on those bits plus 1, and block number r is marked to be retained in Q. Note that the minimum sensible value of an offset is 1.
2. If the two most significant bits of the byte are 01, then the remaining six bits of the byte are prepended (on the left) to the next byte and the two bytes form together a 14-bit (forward) offset from r minus 1.
3. If the most significant bit of the byte is 1, then the remaining seven bits are treated as a bit map, each bit indicating an individual block relative to the value of r. For this purpose, the bits are numbered from 1 to 7 (right to left), and when bit number i is set, the block number $r + i$ is marked to be retained. At the end of processing the byte, r is set to $r + 7$, i.e., the last block number covered by the bit map, regardless of whether the block was marked as retained or not.

The important point is that the interpretation of the consecutive bytes, starting from the front of the packet's payload, produces increasing values of r which eases the operation of updating Q. The queue is scanned in place, in the most natural manner, starting from Q_h, and any blocks whose numbers are not mentioned in the ACK are discarded. Before interpreting the first byte of the ACK, *handle_ACK* initializes r to $L - O_{max} - 1$ to provide a sensible, default, initial value (so the first byte of the ACK can be a forward offset). It might seem natural to initialize r to $L - O_b$ (based on the value of O_b passed in the EOT packet), which would make the range of the initial offset even better contained. However, while the value of L is nailed to the last train received by the Peg, O_b may change in the different (retransmitted) versions of the EOT packet for the same train. As the Tag cannot know which particular copy of EOT served the Peg as the basis for its ACK, O_b is not a well-known value that both parties can always agree on.

Following the reception of an ACK packet from the Peg, the transmitter thread will start the next train with a trimmed-down version of Q. The queue will have been emptied of all blocks that (1) have numbers less than or equal to L from the previous train, and (2) have *not* been mentioned in the ACK packet.

8.4.4 The Peg

The device passes the received blocks to the computer, locally keeping track of the holes in the received sequence, down to the maximum negative offset from the end of the last train. The blocks are tallied up in a bit map whose fixed size covers the interval O_{max}. For the ease of calculations, the actual momentary coverage of the map is described by the current *base* block number B which is shifted to the newest value of $L - O_b$ learned by the Peg. As B is updated, the (logical) beginning of the bit map shifts automatically (in a circular fashion), so the bit map itself is not shifted (no copying is involved), except for clearing the obsolete entries at the tail.

Having received an EOT packet, the Peg updates the base of its bit map to $L - O_b$, sets its reference block number r to $L - O_{max} - 1$, and begins constructing the consecutive bytes of the ACK packet. Given the next missing block to be accounted for, there are these possibilities which are greedily examined in this order: (1) the last entry in the ACK packet is a bit map byte and the block number falls under its coverage, (2) the block number is within a bit-map range from the current value of r, (3) the block number can be represented by a short offset from r, (4) a long (two byte) offset is needed. In the first case, the block is simply added to the bit map byte without extending the ACK packet. In the second case, if the difference between the block number and r is less than 7, a new bit map byte is added to the packet. Note that a bit map byte comes at the same storage expense as a short offset from r, so there is no point in looking ahead: the greedy approach works fine in the sense that it results in the minimum achievable ACK length. Preference is given to the bit map when, based on the current block number alone, the map stands a chance of accommodating at least one more block.

In the unlikely case when the ACK packet becomes filled up to the limit of its size, the receiving Tag will assume that all the blocks falling behind the last block number represented in the ACK are implicitly marked as missing. Note that this may only happen under highly abnormal conditions where the streaming session is probably broken (see Sect. 8.5.2), and pessimistic assumptions regarding the unknown can do no more harm. We have never witnessed an ACK overflow in our experiments.

8.4.5 Data Packaging

As presented so far, the scheme deals with blocks of (in principle arbitrary) data, and one problem that remains to be addressed is the way of packaging the sensor readings into those blocks. The practical limitation of the block size is ca. 50 payload bytes. Although longer blocks are in principle possible with CC1350, there exist good reasons for limiting the block size to a ballpark of this number [18].

A natural option to consider when dealing with bandwidth limitations is data compression. One can suspect that acceleration readings representing an animal's motion will tend to be redundant at those moments when the animal happens to be relaxed. For the project at hand, one of the goals is to determine the useful rate at which the sensor readings should be collected to provide reliable data for classification. Therefore, our objective is to be able to collect data at the maximum possible rate, preferably without sacrificing its accuracy on lossy compression. In fact, the goal of the large project can be interpreted as implementing an extreme data compression scheme where the accelerometer input is transformed into a compact set of identifiers signaling a few very specific conditions. The present device will serve to amass data for building the requisite classifier, and it seems natural to postulate that that data should not be overly lossy to start with. While dense continued readings of IMU sensors are amenable to potentially efficient and effective lossy compression techniques, e.g., of the type discussed in [25], such techniques require prior assumptions about the nature of the classification algorithms fed by the compressed data.

Regarding loss-less compression, the problem is aggravated by three facts: (1) the amount of individual data (the size of a single reading) is small, (2) the size of a data packet is small, and (3) the packets should be independent because they can be independently lost or re-sequenced. In consequence, sophisticated loss-less compression schemes, e.g., based on dictionaries or predictors [20], are not going to be useful because, for a short packet, the size of the metadata needed by the compression algorithm will outweigh the size of the data proper. Also, an obvious idea like run-length encoding, intended to effectively compress data corresponding to periods of low mobility, is not going to work with the accelerometer because (1) the ADC (Analog to Digital Conversion) conversion process in the sensor is tainted with inherent noise (two consecutive readings are almost never exactly the same), (2) the most relevant fragments of the streamed data (the ones likely to be needed for accurate classification) are unlikely to be reducible by this kind of treatment.

Considering the limitations on the packet size, any inflation of the metadata (incurred by compression) will in fact inflate the bandwidth needed to carry useful data as opposed to reducing it. Even the need to accommodate a variable number of readings in a packet will translate into an additional field required to specify the starting reading number. Combined with the fact that a single acceleration reading amounts to three coordinates (that may be subjected to different motion patterns and compression opportunities), that makes it unrealistic to hope for significant reductions with any kind of loss-less compression.

Instead, we focused on investigating the actual accuracy of the sensor readings and the relevance/significance of the different bits in their binary representations. As it turns out, some of those bits mostly carry noise and can thus be ignored with no impact on the accuracy.

The sensor used in our contraption, MPU9250, is a very typical representative of modern IMU devices. Each of the three elements of the acceleration vector is returned as a 2's complement (signed) 16-bit value. This value is interpreted as a fraction in $[-1.0, 1.0 - 2^{-16}]$ applied as a multiplier to the assumed range of the sensor. The latter can be configured into one of four options: 2 g, 4 g, 8 g, and 16 g, with the first option being the obvious choice for our application.[4]

The sensor has a few more configuration parameters of which the only one relevant for our purpose is the LPF (Low Pass Filer) setting intended to remove high frequency components from the signal. The option covers 7 levels corresponding to LPF frequencies from 5 to 460 Hz.

To determine the significance of the binary digits of the acceleration readings, we have carried out a number of experiments where the sensor was kept motionless in several different orientations. The acceleration along each axis was perfectly constant, barring imperceptible noise in the rural and desolate setting of our lab. Independently for each of the three coordinates of the acceleration vector we looked at the values assumed by the least significant bits trying to assess their stability. In the best of the possible worlds, assuming the perfect precision of the readouts, all those bits should always return the same (unchanging) values.

To estimate the amount of noise in the least significant bit of a reading one can simply count the number of times that bit reads as a 0 or 1. But the flips of more significant bits may be triggered by noise in less significant bits due to the carry phenomenon. Thus, the proper way of looking at the significance of k bits is to calculate their combined entropy as:

$$E(k) = -\sum_{i=0}^{2^k-1} F_i \times \log_2(F_i) \tag{8.1}$$

where F_i denotes the frequency of value i taken by those bits, which is between 0 and $2^k - 1$.[5] This way, $E(k)$ comes out as a number between 0 and k. The specific entropy of a given bit m is then calculated as $H_m = E(m) - E(m-1)$, assuming that $E(0) = 0$, with 1 indicating pure randomness, and zero corresponding to a perfectly stable unchanging value.

In Table 8.1 we list the results from four representative experiments. Every row represents a single coordinate, with three consecutive rows amounting to one complete experiment carried out for some specific position of the sensor and some specific setting of the LPF. The sensor position is described by the numbers in the *reference* column which show the average (in principle fixed) values for each of the three coordinates. Eight least significant bits of every coordinate are considered, numbered 1 through 8, starting from the least significant one. The entry in each column shows the value of H_m for the given bit.

For a low bandwidth setting, the four least significant bits primarily carry noise. Bits 5 and 6 appear to be somewhat significant, although their relevance is questionable, to

---

[4] This is because acceleration in excess of 2 g is practically impossible when the sensor is attached to a dog's collar.

[5] In the case when $F_i = 0$ the product under the sum is assumed to be zero.

Table 8.1 The measured entropy of low bits of acceleration readings

Bandwidth	Reference	$k=1$	$k=2$	$k=3$	$k=4$	$k=5$	$k=6$	$k=7$	$k=8$
5 Hz	0.000	1.000	1.000	1.000	1.000	0.980	0.718	0.010	0.000
	−0.160	1.000	1.000	1.000	0.999	0.985	0.874	0.137	0.000
	0.500	1.000	1.000	0.999	0.999	0.998	0.878	0.236	0.001
5 Hz	−0.300	1.000	1.000	1.000	1.000	0.905	0.480	0.018	0.000
	−0.400	1.000	1.000	1.000	0.999	0.754	0.083	0.000	0.000
	−0.040	1.000	1.000	1.000	1.000	0.999	0.855	0.172	0.000
184 Hz	−0.330	0.000	0.000	1.000	1.000	0.905	0.480	0.259	0.002
	−0.150	0.000	0.000	1.000	0.999	0.776	0.303	0.160	0.003
	−0.370	0.000	0.000	1.000	1.000	0.888	0.370	0.201	0.001
460 Hz	−0.076	0.000	0.000	0.000	1.000	1.000	0.922	0.309	0.002
	0.400	0.000	0.000	0.000	1.000	1.000	0.928	0.289	0.001
	−0.010	0.000	0.000	0.000	1.000	0.999	0.996	0.818	0.129

say the least. It appears to improve somewhat for larger magnitudes of the reading. Note that at the range setting of ±2 g, the ubiquitous acceleration of 1 g corresponds to ±0.5.

The significance becomes worse as the bandwidth increases, which is not surprising. At high bandwidth settings the lowest bits are turned off by design; these are bits 1 and 2 for 184 Hz, and 1–3 for 460 Hz. Their entropy is listed as 0 because they are unused by the ADC and stay permanently set at zero regardless of the position of the sensor. Significance picks up at bit 7 with bit 8 still exhibiting a considerable amount of noise at 460 Hz.

The results rather clearly indicate that the six least significant bits of a reading consist dominantly of noise, so ignoring them is not going to impact the accuracy of the collected data. Our targeted collection rate is of order 100+ samples per second which gets us into the high bandwidth region where the accuracy additionally deteriorates. Consequently, we have decided to truncate the coordinate representation to the most significant 10 bits which, as demonstrated in Table 8.1, provide for all the significance that can be squeezed out of the sensor.

The assumed layout of a data block is shown in Fig. 8.5. The 50-byte payload is filled with 12 samples, with the three coordinates of each sample stored in 4 consecutive bytes (a 32-bit word). To avoid wasting two bits per every four bytes, the 24 bits unused by the samples are recycled to store the most significant three bytes of the block number sliced into 12 two-bit fragments. The first two bytes of the block

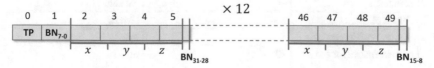

Fig. 8.5 The layout of a data block

store the packet type and the least significant byte of the block number. Not a single bit of the packet payload is left unused.

8.5 Performance

The first implementation of our scheme addresses a planned series of experiments with animals aimed at collecting 128 acceleration samples per second. Our intention was to tune the parameters of the protocol until we would get a satisfying performance for the task at hand. Table 8.2 lists the numerical values of those parameters assumed for the initial tests.

While some of the values in Table 8.2 may be treated as accidental, they were produced by confronting our expectations with the parameters and capabilities of the hardware at hand. A packet carrying 12 samples takes 50 bytes of payload (see Sect. 8.4.5), which translates into 58 bytes of the complete frame, including the physical-layer preamble. Its transmission takes about 9.3 ms which, together with the spacing of 5 ms per packet, yields about 14.3 ms per sample block. We shall use this figure to estimate the maximum capacity of the channel assuming that the sample-carrying blocks constitute the only traffic and that there are no errors. For the 50 kbps rate, it translates into about 70 blocks per second, i.e., 840 samples per second. This is what we mean by the *raw bandwidth* as expressed by R in our calculations below.

Our experiments have demonstrated that it is virtually impossible to lose data in a streaming session controlled by our scheme for as long as the session operates within the bracket of raw technical feasibility. We can handle sampling frequencies

Table 8.2 A tentative setting of protocol parameters

Parameter	Value	Units
Targeted sampling rate	128	samples/s
Channel transmission rate	50	kbps
Samples per block	12	3-vectors
Total block length	58	bytes
End of train packet length	18	bytes
Minimum acknowledgment length	10	bytes
Max. Q size: QS_{max}	128	blocks
Train length: T	64	packets
Max. block offset: O_{max}	2047	blocks
Packet space (within train)	5	ms
End of train space (for the ACK)	20	ms

Fig. 8.6 The performance
model

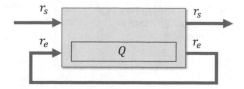

up to 512 samples per second (which is just one notch below the maximum capacity
of the sensor) without increasing the channel rate, and up to the maximum of 1024
samples per second at a slight increase of the channel rate, practically without losing
any samples.

8.5.1   Formal Analysis

Let R denote the maximum rate at which blocks can be transmitted, assuming smooth
operation and no errors. We shall refer to R as the raw available bandwidth of the
channel (see above). Let r_s denote the target effective rate corresponding to the frequency at which we would like to reliably receive blocks of sensor readings. Let
r_e be the block retransmission rate of the channel, i.e., the frequency at which the
blocks that have been transmitted but not received materialize for retransmission.
The system can be modeled as a server shown in Fig. 8.6 with the blocks viewed as
customers. The bottom path represents the traffic incurred by errors and the consequent retransmissions requested by the Peg in its acknowledgments. This is what the
queue Q is intended for: to accommodate blocks that must be retransmitted because
of errors.

Consider the system at equilibrium and note that the upper path is stable and deterministic: the blocks arrive at a steady rate r_s clocked by the sensor, their processing
(transmission) time is fixed, and they leave the server at the same rate. Consequently,
we can ignore the trivial dynamics of the upper part assuming that its impact consists
in removing from R a fixed portion amounting to r_s. Whatever is left, i.e., $R - r_s$
can be treated as the bandwidth available for the bottom part of the traffic, i.e., for
retransmissions.

Suppose that the errors are independent and they occur at the same probability P_e
for every transmitted block. Then we have:

$$r_e = r_s \times \sum_{i=1}^{\infty} P_e^{\,i} = \frac{r_s \times P_e}{1 - P_e} \qquad (8.2)$$

The retransmission service can be modeled as an M/D/1 queue with the utilization parameter $\rho = \lambda T$ expressing the saturation level of the system [30]. In this

Fig. 8.7 $C =$ the average occupancy of Q versus the utilization factor ρ

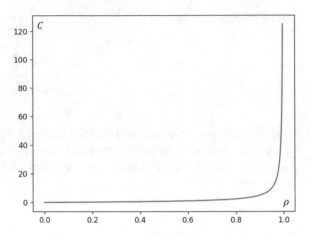

standard formula, λ is the customer arrival rate and T is the service time; thus, the system is considered to be fully saturated when $\rho = 1$. In our model, we shall use the transmission time of a complete block of samples as the time unit. Consequently, we have:

$$\rho \approx \frac{r_e}{R - r_s} \tag{8.3}$$

which can be interpreted as the fraction of the channel bandwidth that remains available for retransmissions (after accounting for the steady arrival rate of samples from the sensor) that is in fact being used by the retransmissions under the present load.

The expected occupancy of the queue in the standard M/D/1 model is [30]:

$$C = \rho + \frac{1}{2} \left(\frac{\rho^2}{1 - \rho} \right) \tag{8.4}$$

The graph of C versus ρ (Fig. 8.7) is illuminating and reassuring. It shows that unless the utilization factor becomes very close to unity, i.e., the retransmissions fill all the bandwidth available to them, the demand for queue space is modest. Consequently, to see the queue overflow and actual losses start to materialize, we have to bring the system to the very edge of its equilibrium.

To be more accurate, we should notice that (8.3) underestimates the value of ρ by ignoring the acknowledgments. Their main impact is in stealing an additional fragment of R proportional to the total portion of the bandwidth used by the trains. To factor it in, we should express the utilization parameter as:

$$\rho = \frac{r_e}{R - r_s - r_a} \tag{8.5}$$

with r_a representing the portion of the bandwidth needed for the acknowledgments. The latter can be written as:

Fig. 8.8 The utilization parameter ρ as a function of P_e for different sampling rates

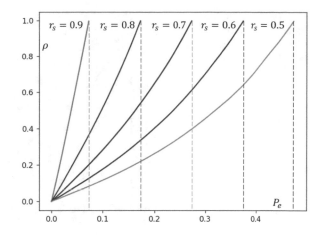

$$r_a = \frac{A \times (r_s + r_e)}{T \times (1 - P_e)} \tag{8.6}$$

where A is the bandwidth needed to expedite a single acknowledgment packet expressed as the ratio of its transmission time to the time required to transmit a block with samples (which is the unit of bandwidth in our model), and T is the number of packets in a train. The $1 - P_e$ factor in the denominator accounts for the possibility that the acknowledgement itself may have to be retransmitted because of an error. Strictly speaking, the (average) size of the acknowledgment packet A is a (rather messy) function of r_e; however, the way the lost samples are represented in the ACK (see Sect. 8.4.3) practically guarantees that under any sensible equilibrium A fluctuates very little, within a few bytes. Note that the operation of acknowledging a train involves a fixed-length EOT packet (issued by the Tag), followed by a 5 ms space, followed in turn by the transmission of the ACK packet by the Peg. Normally all this takes ca. 10 ms which is marginally more than the total cost of transmitting a sample block (estimated at 9.3 ms), yielding $A \approx 1.1$. There is no reason to split hairs and consider the variability in the size of an ACK packet, especially that A only kicks in once per every train-full of packets. In particular, in our experimental setting $A/T \approx 0.017$ which indicates that the impact of A generally tends to be small.[6] Figure 8.8 shows how the utilization parameter changes with the increasing probability of losing a sample block for five different sampling rates r_s, expressed as a fraction of R. The point where ρ reaches 1 indicates the upper bound for the probability of error for the given sampling rate, which we shall denote by P_{e_m}.

The relationship between r_s and P_{e_m} is almost linear, and it would be perfectly so, if not for the impact of the acknowledgments. Indeed, the maximum corresponds to $\rho = 1$ which, from (8.2) and (8.3), implies:

---

[6] This was the point of the proposed scheme, as stated in Sect. 8.3.

Fig. 8.9 The maximum
probability of bit error P_{e_m}
versus the sampling rate r_s
for different values of A

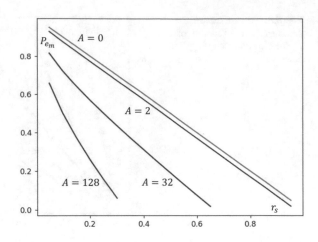

$$P_{e_m} = \frac{R - r_s}{R} \qquad (8.7)$$

This idealistic dependence is disturbed by (8.6) into a quadratic equation whose
solution yields:

$$P_{e_m} = 1 - \frac{r_s + \sqrt{4R \times r_s \times \frac{A}{T} + r_s^2}}{2R} \qquad (8.8)$$

Note that for $A = 0$, i.e., with the acknowledgments ignored, Equation (8.8) boils
down to (8.7).

Figure 8.9 illustrates how P_{e_m} depends on A for $T = 64$. The curve for $A = 2$
corresponds to a slightly exaggerated estimate from our experimental setup.

From the viewpoint of reliability, the most interesting aspect of the behavior of
our scheme is the frequency of overflowing the buffer space (the queue Q) which
will translate into hard losses. Figure 8.7 suggests that loses are going to be rare,
unless ρ becomes very close to 1, but it does not show the distribution of loses for
a realistic queue with finite buffers. For a quantitative insight into this problem we
need to study an M/D/1/N system with a finite queue where an overflowing customer
is discarded. Such a system was investigated in [10] and [17] where the following
formulas were given for the probability P_j of finding the system in the state with j
customers awaiting service:

$$P_0 = \frac{1}{1 + \rho b_{N-1}} \qquad (8.9)$$

$$P_N = 1 - \frac{b_{N-1}}{1 + \rho b_{N-1}} \qquad (8.10)$$

$$P_j = \frac{b_j - b_{j-1}}{1 + \rho b_{N-1}}, \qquad j = 1, \ldots, N - 1 \qquad (8.11)$$

where:

$$b_n = \sum_{k=0}^{n} \frac{(-1)^k}{k!}(n-k)^k e^{(n-k)\rho} \rho^k \qquad (8.12)$$

By setting N to the queue size and calculating the probability of finding N customers in the system, we obtain an upper bound on the probability of losing a block for a given value of ρ.

The proper way of interpreting the queue Q in the context of the M/D/1/N model calls for some discussion. Formally, the model stipulates that the service time is fixed and the customers (i.e., retransmissions) arrive at random according to the Poisson process. In the real system, the retransmissions materialize, possibly in bunches, with the arrival of the ACK packet, which event tends to occur at regular intervals determined by the train length T. When it happens, the retransmissions triggered by the ACK are already queued (present in Q). This may look like a potentially significant departure from the model.

The way we should look at it is that a retransmission formally "arrives" at the server at the moment when a packet transmitted by the Tag fails to be received at the Peg. This is when the customer is conceptually put into the queue of the model. With this interpretation, owing to the steady transmission rate of all queued blocks, the stipulation about the Poisson arrival process of the retransmissions is equivalent to the assumption that transmission errors are independent events occurring at some stable long-term rate. But then, the queue in the real system is still utilized differently than in the model because (1) the retransmissions are pre-queued ahead of their formal arrival time, (2) the service for them may be delayed until the ACK is received by the Tag. Note, however, that the assumption about the fixed service rate for the retransmissions still holds over a longer term, if we compensate for the hiccups resulting from the processing of the trains. The server is obviously work conserving, so whatever bandwidth is left for the retransmissions is not going to be used for anything else. Consequently, a crude but completely safe way to appease the formal model would be to agree that the portion of Q needed to accommodate a full train acts as a filter interfacing the system to the model by absorbing the discrepancies between the two and making them compatible. This boils down to the assumption that the number N of queue slots for the customers in our M/D/1/N model is equal to $QS_{max} - T$.

Even with this sacrifice the numbers come out quite favorably. Figure 8.10 shows the distribution of the probability of finding an M/D/1/N server with $N = QS_{max} - T = 64$ in a state with $k \geq n$ customers present in the system. The last value can be viewed as an upper bound on the probability of rejecting a customer upon arrival, i.e., losing a sample block. Even for $\rho = 0.95$ the probability of a loss is of order 10^{-4}, while being practically negligible for any lesser load.

Fig. 8.10 The probability of finding the system in a state with the number of queued customers $\geq n$, for the queue length $N = 64$

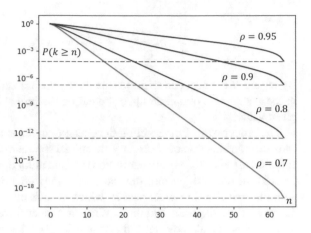

8.5.2 Experiments

We built an experimental data collection system consisting of a CC1350 SensorTag communicating with a CC1350 LaunchPad Peg connected to a PC via a USB cable. Figure 8.1 depicts the SensorTag in the form it arrives from the manufacturer (including the elastic protective jacket), which makes it directly usable for experiments. The devices were programmed in PicOS [18]. The parameters for experimentation were set according to Table 8.2 with the exception of bumping the sampling rate to 256 samples per second to stress the system one notch above its intended load. We were interested in measuring losses under realistic conditions. The received samples were dumped by the Peg to a file as they arrived from the Tag, with the retransmitted blocks stored out of order, and postprocessed off-line into a continuous stream of data.

To monitor the status of the queue Q without incurring additional traffic between the Tag and the Peg, we temporarily replaced a few sensor readings in the sample blocks with the values of selected variables from the Tag. The devices operated with no external antennas. The effective communication range was about 100 m in open field and 60-70 m in our residential setting consisting of a detached home and a yard, which well represented a typical area of practical interest for the project at hand. The range could have been easily extended by equipping the Peg with a better antenna, but our intention was to actually provoke packet losses without overly complicating the test scenarios.

Owing to the extremely low probability of losing samples on an intermittent and accidental transmission error, the system tends to behave in a bipolar manner. For as long as the devices remain within a mutual communication range, where the transmission errors are reasonably random and their rate does not exceed the saturation threshold, there are simply no losses at all, no matter how long one waits. This much can be inferred from Sect. 8.5.1, especially from Fig. 8.10. When the devices are separated beyond some critical point, even so slightly, the system breaks down, the

Time:	14h 17m 22s
Rate:	256
Total blocks:	1097429
Accounted for:	1097429
Out of order:	61
Lost (OSS/Peg):	0
Dropped:	0

Scenario 1

Time:	5h 44m 19s
Rate:	256
Total blocks:	440725
Accounted for:	440725
Out of order:	677
Lost (OSS/Peg):	0
Dropped:	0

Scenario 2

Fig. 8.11 Two healthy scenarios with zero losses

queue overflows, and the losses become intolerable. It is virtually impossible to adjust the separation to see steady losses amounting to some middle-point average. Note that by losses we mean here samples missing for good, as opposed to occasional reception errors from which the system is able to recover.

Figure 8.11 lists the summaries of two experiments corresponding to what we we would call "healthy" scenarios, with the second one intended to represent typical usage. In the first case, the two devices were placed at fixed positions in a room separated by a distance of ca. 3 m, started, and left alone overnight. The second experiment involved a dog wearing the Tag on the collar (Fig. 8.1) and wandering, in the course of its more or less standard daily routine, within the perimeter of ca. 60 m occasionally taking a nap.

In all runs, when the sampling process at the Tag was stopped, the device was allowed to send whatever backlog of blocks remained in the queue until all of them were acknowledged or became obsolete (see Sect. 8.4.1). The total number of blocks meant to arrive at the Peg was determined from the last-received EOT packet. The "accounted for" field lists the number of blocks that (eventually) made it to the Peg, with "lost" being the difference between the two. These numbers were calculated by the postprocessor of the data dumped by the Peg after the run was terminated. While combining the blocks into a continuous sequence of samples, the postprocessor also counted the blocks that arrived "out of order," i.e., ones that had to be retransmitted. An out-of-sequence block was only counted once, regardless of how many times it might have been retransmitted. Thus, the ratio of that number to the total (expected) number of blocks gives the measured average block error rate, which was 5.6×10^{-5} and 1.5×10^{-3}, respectively, in each of the two scenarios. The "dropped" field lists the number of packets that had to be discarded from Q before being acknowledged. This is the number of lost blocks as perceived by the Tag. Note that it does not have to perfectly match the number of blocks deemed lost at the final collection point. First, the failure of an acknowledgment to materialize before the block is discarded does not have to mean that the block has not been received by the Peg (the ACK may have been lost or delayed). Second, it is technically possible (although unlikely and never witnessed) that a packet is lost in the serial-over-USB connection of the Peg to the processing host. For the purpose of our experiments that link was not made formally foolproof, although it can be easily made so.

```
Time:              0h 5m 24s          Time:              0h 3m 2s
Rate:              256                Rate:              256
Total blocks:      6791               Total blocks:      3801
Accounted for:     5518               Accounted for:     804
Out of order:      2818               Out of order:      677
Lost (OSS/Peg):    1273               Lost (OSS/Peg):    2997
Dropped:           1273               Dropped:           2997
```

                    Scenario 3   Scenario 4

Fig. 8.12 Two unhealthy scenarios with serious losses

To see the relevance of the queue Q, beyond the obviously needed portion of size up to T accommodating the current train, we included a report of its momentary occupancy in every outgoing block of samples. We were only interested in those cases when the queue was filled to a level above $T = 64$, i.e., in excess of the size of a full train which Q must accommodate in any normal course of events. The maximum excess ever noted in the scenarios shown in Fig. 8.11 was 1, which demonstrates that the sacrifice made in Sect. 8.5.1 to eliminate the discrepancy between the model and the real system was exaggerated. Having reached the occupancy of T (at the end of the current train), the queue is not going to discard a block until the Tag receives an ACK from the Peg. Normally, the procedure takes about 10 ms (see Sect. 8.5.1). The time interval until the next block of samples becomes filled up and ready to be added to the queue is $12/256\,\mathrm{s} \approx 47$ ms. This allows for at least four (practically about 6) full retries (EOT-ACK); thus, even if the first ACK does not make it, the chance that the queue is going to remain filled up to the size of T when the next block needs to be stored is very small. While this may not be completely impossible under conditions of heavy losses (also affecting the acknowledgments), what is in fact almost impossible is that the ACK will still fail to make it to the Peg within twice that interval (which would push the occupancy of Q over $T + 1$). Once the ACK arrives, a few slots in the queue are bound to be vacated (even under extreme losses), thus becoming a useful portion of the service queue for new retransmissions. This means that the effective value of N for the model (see Sect. 8.5.1) is larger than $QS_{max} - T$.

For the record, Fig. 8.12 shows two cases involving hopeless losses with the devices operating beyond the saturation threshold. The experiments only lasted for a few minutes because it immediately became clear that continuing them was pointless.

Setting up a real-life environment where RF interference naturally follows scripted scenarios is extremely difficult. Thus, as part of the testbed for our implementation, we modified the code at the two devices to simulate randomized errors occurring according to prescribed parametrizable schemes. In one series of experiments, the devices were placed on a table at a minimal separation distance (about 1.5 m) in a sterile RF environment, thus practically precluding natural losses, and set to drop the received packets randomly with the probability proportional to their length, as to model a steady and independent bit error rate P_b. More precisely, assuming a given value of P_b, we made the probability of dropping a packet of length l (in bytes) equal

Time:	1h 0m 33s
Rate:	256
Total blocks:	77501
Accounted for:	77501
Out of order:	68946
Lost (OSS/Peg):	0
Dropped:	0

Time:	1h 0m 23s
Rate:	256
Total blocks:	77285
Accounted for:	75282
Out of order:	67165
Lost (OSS/Peg):	2003
Dropped:	2005

Scenario 5   Scenario 6

Fig. 8.13 Two scenarios with artificially induced losses

to $P_b(l) = 1 - (1 - P_b)^{8l}$. The block error rate used in Sect. 8.5.1 can be expressed as $P_e = P_b(58)$. By adjusting the value of P_b (by bisection) we were able to pinpoint the threshold of losses at which the system would break down, equivalent to P_{e_m} from Eq. (8.8). Figure 8.13 shows two scenarios from those experiments. In Scenario 5, P_b is set to 0.00188 which was determined as the maximum value for which a 1 h run resulted in zero losses. Scenario 6 has been enacted for a marginally larger value of $P_b = 0.00189$ (we only considered three significant digits) demonstrating the abruptness of the boundary. When we calculate P_{e_m} from Eq. (8.8) for our case, assuming $R = 70$ blocks per second, $r_s = 22$ blocks per second, $A/T = 0.02$, we obtain $P_{e_m} \approx 0.6$ which well matches the observed threshold.

One potentially malicious scenario for our protocol would occur for bursty errors in the RF channel, with the maximum burst duration reaching the capacity of Q. For the sampling rate of 256 samples per second and $Q = 128$, the burst would have to last for $128 \times 12/256 = 6$ s (and amount to total jamming) to cause a guaranteed loss. Owing to the aggressive positive impact of any available queue space on the reliability of delivery, any burst shorter than that would be unlikely to cause any damage at all, provided that the long term average error rate would keep the system below saturation. Note that the way acknowledgment packets are formed (Sects. 8.4.3 and 8.4.4) facilitates compact representation of burst losses as bit maps. We carried out tests with simulated blackouts, produced by logically turning off the Peg's radio (both ends) for prescribed intervals, which confirmed these observations.

8.6   Summary

We have presented a protocol for reliable streaming of telemetric data over an unreliable wireless channel. Our scheme seems to make a good use of the available bandwidth, especially in the specific context of its inspiring application. On the sender's side, this is accomplished by an efficient organization of storage for the outstanding (unacknowledged) packets. The feedback sent by the recipient is minimized to reduce the impact of channel reversals on the bandwidth available for the forward traffic. The proposed scheme is intended for small-footprint wireless sensing devices where the amount of memory for packet buffers is very small. For as long as the quality of the RF channel prevents unrealistically long error bursts, and the long term average error rate remains below the limits of the raw formal capability of the system to deliver all data, our scheme practically guarantees zero losses.

One point implicitly made in our paper is that tiny embedded systems aimed at specific applications within the realm of the so-called Internet of Things (IoT) come out best when built following a "holistic" approach by which we mean taking into account, from the very bottom, the idiosyncratic aspect of the application, instead of relying on ready, layered, standardized, library solutions. In addition to reducing the footprint of the application, and enabling it to run in a cheaper and resource-frugal device, such an approach also translates into better performance, and, with the right selection of tools [18], may in fact speed up the development process while resulting in a better quality of the product.

References

1. Alcaidinho, J.M.: The internet of living things: enabling increased information flow in dog-human interactions. Ph.D. thesis, Georgia Institute of Technology (2017)
2. Baransel, C., Dobosiewicz, W., Gburzyński, P.: Routing in multi-hop switching networks: Gbps challenge. IEEE Netw. Mag. **9**(3), 38–61 (1995)
3. Bartlett, K.A., Scantlebury, R.A., Wilkinson, P.T.: A note on reliable full-duplex transmission over half-duplex links. Commun. ACM **12**(5), 260–261 (1969)
4. Berns, G.S., Brooks, A.M., Spivak, M., Levy, K.: Functional MRI in awake dogs predicts suitability for assistance work. Sci. Rep. **7**, 43704 (2017)
5. Bluetooth SIG: Bluetooth technology (2020)
6. Boruta, A., Gburzyński, P., Kuźnicka, E.: On reliable wireless streaming or real-time sensor data. In: Proceedings of CS&P, pp. 131–144. Berlin, Germany (2021). http://ceur-ws.org/Vol-2951/paper14.pdf
7. Bray, E.E., Levy, K.M., Kennedy, B.S., Duffy, D.L., Serpell, J.A., MacLean, E.L.: Predictive models of assistance dog training outcomes using the canine behavioral assessment and research questionnaire and a standardized temperament evaluation. Front. Veterin. Sci. **6**, 49 (2019)
8. Brložnik, M., Avbelj, V.: A case report of long-term wireless electrocardiographic monitoring in a dog with dilated cardiomyopathy. In: 2017 40th International Convention on Information and Communication Technology, Electronics and Microelectronics (MIPRO), pp. 303–307. IEEE (2017)
9. Brložnik, M., Likar, Š., Krvavica, A., Avbelj, V., Domanjko Petrič, A.: Wireless body sensor for electrocardiographic monitoring in dogs and cats. J. Small Anim. Pract. **60**(4), 223–230 (2019)
10. Brun, O., Garcia, J.M.: Analytical solution of finite capacity M/D/1 queues. J. Appl. Probab. **37**(4), 1092–1098 (2000)
11. Byrne, C., Zuerndorfer, J., Freil, L., Han, X., Sirolly, A., Cilliland, S., Starner, T., Jackson, M.: Predicting the suitability of service animals using instrumented dog toys. In: Proceedings of the ACM on Interactive, Mobile, Wearable and Ubiquitous Technologies, vol. 1(4), pp. 1–20 (2018)
12. Cooke, B.J., Hill, L.B., Farrington, D.P., Bales, W.D.: A beastly bargain: a cost-benefit analysis of prison-based dog-training programs in Florida. Prison J. **101**(3), 239–261 (2021)
13. Duerr, F.M., Pauls, A., Kawcak, C., Haussler, K.K., Bertocci, G., Moorman, V., King, M.: Evaluation of inertial measurement units as a novel method for kinematic gait evaluation in dogs. Veterin. Compar. Orthopaed. Traumatol. **29**(06), 475–483 (2016)

14. Foster, M., Brugarolas, R., Walker, K., Mealin, S., Cleghern, Z., Yuschak, S., Clark, J.C., Adin, D., Russenberger, J., Gruen, M., et al.: Preliminary evaluation of a wearable sensor system for heart rate assessment in guide dog puppies. IEEE Sens. J. **20**(16), 9449–9459 (2020)
15. Foster, M., Mealin, S., Gruen, M., Roberts, D.L., Bozkurt, A.: Preliminary evaluation of a wearable sensor system for assessment of heart rate, heart rate variability, and activity level in working dogs. In: 2019 IEEE SENSORS, pp. 1–4. IEEE (2019)
16. Foster, M., Wang, J., Williams, E., Roberts, D.L., Bozkurt, A.: Inertial measurement based heart and respiration rate estimation of dogs during sleep for welfare monitoring. In: Proceedings of the Seventh International Conference on Animal-Computer Interaction, pp. 1–6 (2020)
17. Garcia, J.M., Brun, O., Gauchard, D.: Transient analytical solution of M/D/1/N queues. J. Appl. Probab. **39**(4), 853–864 (2002)
18. Gburzyński, P.: A WSN architecture for building resilient, reactive, and secure wireless sensing systems. Task Quarterly **25**(2), 141–181 (2021)
19. Gburzynski, P., Kaminska, B., Rahman, A.: On reliable transmission of data over simple wireless channels. J. Comput. Syst., Netw., Commun. **2009** (2009). https://doi.org/10.1155/2009/409853
20. Gopinath, A., Ravisankar, M.: Comparison of lossless data compression techniques. In: 2020 International Conference on Inventive Computation Technologies (ICICT), pp. 628–633. IEEE (2020)
21. Jenkins, G.J., Hakim, C.H., Yang, N.N., Yao, G., Duan, D.: Automatic characterization of stride parameters in canines with a single wearable inertial sensor. PLoS ONE **13**(6), e0198893 (2018)
22. Kurose, J.F., Ross, K.W.: Computer Networking: A Top-Down Approach Featuring the Internet. Addison-Wesley (2004)
23. Kuźnicka, E., Gburzyński, P.: Automatic detection of suckling events in lamb through accelerometer data classification. Comput. Electron. Agric. **138**, 137–147 (2017)
24. Lakshman, T., Madhow, U.: The performance of TCP/IP for networks with high bandwidth-delay products and random loss. IEEE/ACM Trans. Netw. **5**(3), 336–350 (1997)
25. Lin, J.W., Liao, S.w., Leu, F.Y.: Sensor data compression using bounded error piecewise linear approximation with resolution reduction. Energies **12**(13), 2523 (2019)
26. Lin, S., Costello, D., Miller, M.: Automatic-repeat-request error-control schemes. IEEE Commun. Mag. **22**(12), 5–17 (1984)
27. Lippi, G., Cervellin, G., Dondi, M., Targher, G.: Hypoglycemia alert dogs: a novel, costeffective approach for diabetes monitoring? Altern. Ther. Health Med. **22**(6), 14 (2016)
28. Lynch, W.: Reliable full-duplex transmission over half-duplex telephone lines. Commun. ACM **11**(6), 407–410 (1968)
29. Nafaa, A., Taleb, T., Murphy, L.: Forward error correction strategies for media streaming over wireless networks. IEEE Commun. Mag. **46**(1), 72–79 (2008)
30. Nakagawa, K.: On the series expansion for the stationary probabilities of an M/D/1 queue. J. Oper. Res. Soc. Japan **48**(2), 111–122 (2005)
31. Pereira, R., Pereira, E.G.: Video streaming considerations for internet of things. In: 2014 International Conference on Future Internet of Things and Cloud, pp. 48–52. IEEE (2014)
32. Schoenfeld-Tacher, R., Hellyer, P., Cheung, L., Kogan, L.: Public perceptions of service dogs, emotional support dogs, and therapy dogs. Int. J. Environ. Res. Public Health **14**(6), 642 (2017)
33. Slabbert, J.M., Odendaal, J.S.: Early prediction of adult police dog efficiency–a longitudinal study. Appl. Anim. Behav. Sci. **64**(4), 269–288 (1999)
34. Texas Instruments: CC1350 SimpleLink Ultra-Low-Power Dual-Band Wireless MCU (2019). http://www.ti.com/lit/ds/symlink/cc1350.pdf. Technical document SWRS183B
35. Yount, R.A., Olmert, M.D., Lee, M.R.: Service dog training program for treatment of posttraumatic stress in service members. US Army Med. Depart. J. **8**(4/5/6), 63–69 (2012)

Chapter 9
Graph-Based Sparse Neural Networks for Traffic Signal Optimization

Łukasz Skowronek, Paweł Gora⊙, Marcin Możejko⊙, and Arkadiusz Klemenko⊙

Abstract We investigate the performance of sparsely connected neural networks, with connectivity determined by road network graphs, for solving the Traffic Signal Setting optimization problem. We conducted experiments on three realistic road network topologies and found these types of graph neural networks superior to fully connected ones, both in terms of generalization properties on fixed test sets and—more importantly—near target function minima obtained in the gradient descent optimization process. We additionally confirm the soundness of our method by showing that random perturbations of the actual graph lead to consistent deterioration of model performance.

9.1 Introduction

Traffic optimization problems have a natural underlying graph structure, determined by the topology of the corresponding road network. In this paper, we introduce a neural network architecture based on a road network graph adjacency matrix to solve the so-called Traffic Signal Setting (TSS) problem, in which the goal is to find the optimal traffic signal settings for given traffic conditions (as defined in [5]):

Definition 9.1 Traffic Signal Setting (TSS) problem:

- Given is a directed graph of a road network with traffic signals located in some vertices. Traffic signals are objects with attributes: duration of a red signal phase

Ł. Skowronek
Brygadzistów 23a/3, 40-807 Katowice, Poland

P. Gora (✉) · M. Możejko
Institute of Informatics, Faculty of Mathematics, Informatics and Mechanics, University of Warsaw, Banacha 2, 02-097 Warsaw, Poland
e-mail: p.gora@mimuw.edu.pl

A. Klemenko
Garbary 95B/95, 61-757 Poznań, Poland

© The Author(s), under exclusive license to Springer Nature Switzerland AG 2023
B. Schlingloff et al. (eds.), *Concurrency, Specification and Programming, Studies in Computational Intelligence 1091*, https://doi.org/10.1007/978-3-031-26651-5_9

(TR), duration of a green signal phase (TG), offset (TS)—values of these attributes may be modified.

- **Traffic Signal Setting (TSS)**—set of values (TG, TR, TS) for all signals in a road network.
- Given is a virtual traffic model: cars with initial speeds, positions in some vertices of the road network graph, static routes, and rules of drive on edges.
- Given is an objective function F which calculates the quality of a traffic signal setting.
- Goal: Find a traffic signal setting for which the value of F is (sub)optimal.

Some variants of this problem were proven to be NP-hard even for very simple traffic models [24], and therefore, heuristics and approximations were used to solve it [5], but the existing approaches still have some drawbacks. For example, evaluating the quality of traffic signal settings using accurate traffic simulations (which is a standard evaluation method) can be too time-consuming, especially for large-scale road networks and/or online evaluation [6, 14]. Also, the size of the space of possible solutions is so large that it turns out infeasible, in any reasonable time, to obtain global minima (or even relatively good signal settings) of the simulator output by checking all the possible solutions or doing a random search, as most points in the input space are far from the optimal solutions [5].

A strategy used to overcome these two difficulties was presented in [5] and consists of generating a reasonably sized training set using a traffic simulator and then fitting a machine learning model to approximate the outcomes of traffic simulations very fast and accurately. Then, an optimization algorithm, such as gradient descent or genetic algorithm, can be applied to find traffic signal settings that are heuristically optimal according to the trained metamodel. Such settings can be later evaluated using the traffic simulator that was used to generate the training set, in order to validate if the settings considered as good by the metamodel are also good according to the simulator. This strategy turned out to be quite successful [4, 5, 14], yet the models' accuracy degraded close to points considered as minima by the optimization algorithm and the model, making further optimization far more difficult [6, 14].

In this paper, we introduced graph-based neural networks (GNN) that can outperform feed-forward fully connected neural networks (FCNN) on the task of approximating traffic simulation outcomes. Compared to the standard FCNN, the introduced GNN have most of the connections deleted, keeping only the crucial connections between neurons (making the information flow corresponding to the traffic flow in the road network), which makes this architecture relatively sparse, easier to train and generalize. As a consequence, GNN have better accuracy on the test set, as well as close to local optima found using the gradient descent optimization applied to the TSS problem.

We also convincingly prove that the GNN architecture works better than analogous ones based on unrelated or perturbed adjacency matrices, as well as—crucially—better than a number of simple FCNN architectures that also perform reasonably well for the problem. There are several possible advantages of using sparsely connected, graph-based neural networks. In this work, we focus on generalization properties, in which the graph structure can be considered as acting as a regularizer. We plan to investigate other possible advantages of GNN (like better scalability and parallelizability, faster convergence, better visualization possibilities) in the future. It is also important to emphasize that even though we describe an application of the considered GNN to a specific problem (TSS), we can imagine their usage in many contexts outside the traffic optimization domain. Wherever a problem has a clear spacial/neighbourhood structure, GNN can be tried as a feasible network regularization technique.

The rest of the paper is organized as follows. Section 9.2 puts our work in the context of a research in the domain of building surrogate models for complex processes, solving TSS problem and using graph neural networks. Section 9.3 presents the two types of graph neural network architectures we used. In Sect. 9.4, we describe the setup of our main experiments including a description of the used datasets and the process of their generation. Section 9.5 summarizes our main experiment results showing that graph neural network architectures introduced in this paper outperform other models used in such a task. In Sect. 9.6, we summarize the results of several 'sanity checks' we performed in order to confirm that our results were not obtained by chance. Moreover, the results in Sect. 9.6 can be especially interesting for researchers in graph neural networks. For example, we show that out-of-sample performance of graph-based sparse neural nets decreases (almost) monotonously as a function of the distance of the adjacency matrix that we use for constructing the network from the true adjacency matrix. We summarize the presented research in Sect. 9.7, outlining some possible future research directions.

9.2 Related Works

Complex processes, such as road traffic in cities, are difficult to study, due to large number of interacting components (e.g., vehicles), nondeterminism or sensitive dependence on initial conditions. Very often, the only reasonable method to accurately predict the behaviour of such systems is to apply computer simulations which can be time-consuming and usually can't be simplified due to computational irreducibility. However, in many tasks related to complex processes it is not necessary to obtain a very accurate prediction, it could be sufficient to get only an approximate outcome but achieve it as fast as possible (due to stochasticity or sensitive dependence on initial conditions, it may be impossible to predict the exact value anyway). Therefore, in such cases it is natural to try to build the so-called surrogate models (metamodels) approximating outcomes of simulations very fast and with good accuracy [29]. Such applications are especially common in the case of optimization tasks, in which it is often necessary to run multiple simulations in order to evaluate many different input settings [8, 9, 18]. This is the case of the traffic optimization problems [17], such as TSS problem.

Many such surrogate models are based on machine learning methods, such as neural networks [9, 18], and also some previous works on solving TSS [4–6, 14] use various machine learning techniques (mainly based on neural networks and gradient boosted decision trees) to build metamodels of traffic simulations that were used to evaluate quality of different traffic signal settings. Such surrogate models were able to approximate outputs of simulations (the total times of waiting on red signals) with very good accuracy (e.g., values of the mean absolute percentage error (MAPE) metrics were at the level 1−2%) and a few orders of magnitude faster than by running microscopic simulations [4, 5]. Thanks to that, it was possible to use optimization algorithms such as genetic algorithms, simulated annealing, Bayesian optimization or gradient descent, to find heuristically optimal traffic signal settings without performing extensive parameter space searches that would take weeks to complete [4–6, 14]. However, information about the road network structure was never used in those experiments, even though it should be important when optimizing traffic.

In this paper, we fill in the gap by applying a family of sparsely connected neural nets, with connectivity determined by the adjacency matrix of a road network graph, to the TSS problem.

Introducing the direct connection between the network architecture and the graph structure helps to leverage additional information represented by a graph. Similarly to our work, [13] introduces a graph neural network layer in which each vertex has some specific parameters assigned to combine information from its neighbors. However, differently to our method, this layer uses only an original graph matrix and skips the dual graph structure when performing computations. The only notable usage of a dual structure we found is in [1], where it is compressed to a matrix using aggregated statistics from a random graph walk procedure. This aggregation is used to introduce a vertex neighborhood context similarly to a popular T-SNE method [27]. An extensive overview of different graph neural network architectures and applications can be found in [28].

Due to their capability to capture a road network structure, GNN were used in multiple traffic applications. In [7, 12, 26] authors used spatio-temporal GNN for traffic situation prediction, whereas [25] used the same technique in order to predict the TAXI demand. However, our application of graph neural networks in the traffic optimization domain and to the TSS problem seems to be the first such approach.

9.3   Network Architecture

The key idea in defining our sparse graph-based neural network architecture is an intuitively compelling rule that information/signal should propagate locally between the net layers. By locality, we mean the presence of only those neuron connections that have a corresponding non-zero entry in the adjacency matrix of the corresponding graph. In the case of the road network, in order to implement such a rule, the neurons in the successive layers of the neural network should be linked to the neurons corresponding to vertices and/or edges of the corresponding graph. Thus, we propose the following general ways to build a graph neural network:

1. Neurons in the even numbered layers, starting with the input layer as layer 0, correspond to graph vertices (in our case—road crossings). Neurons in the odd numbered layers correspond to graph edges (in our case—road segments). An exception should be the final layer with just one neuron. Connections from a vertex-localized layer to an edge-localized layer should only be present if a given vertex is an end of a given edge in the corresponding road network graph. There are exactly two such connections for every edge neuron. Connections from an edge-localized layer to a vertex-localized layer should only be present if the edge has the vertex as its end in the corresponding road network graph. The number of such connections is equal to the number of particular vertex neighbors.

or

2. Neurons in all layers, with the exception of the output layer, correspond to road network graph vertices. Connections from a neuron in one layer to a neuron in the next one should only be present if the corresponding vertices are neighbors in the road network graph. The number of connections for the vertex node is equal to the number of the vertex neighbors.

Although architecture 2 might seem to be more basic, architecture 1 appears to naturally model a traffic flow through the road networks (see Sect. 9.3.1 for a detailed explanation). In the rest of this paper, we focus solely on GNN of the architecture type 1.

It should also be pointed out that GNN can have multiple channels at each edge/vertex. In the following part, we always assume the number of channels is constant across the hidden layers of the network, and is a hyperparameter of the network.

One may also notice a similarity between our GNN architecture of type 2 and the graph neural networks proposed by Thomas Kipf [11]. However, we do not share any weights in our model, as we aim to focus on local patterns connected to roads/crossroads. Theoretically, we could introduce some weight sharing in the 'edge' layers of GNN of type 1, but our first experiments using this approach led to highly disappointing results.

In typical machine learning literature terminology, our GNN should likely be called 'neural networks with a fixed sparse connectivity mask'. In the case of multi-channel networks, sparsity is applied in the 'spacial', but not in the 'channel' dimension.

9.3.1 Graph Neural Network Architecture—Detailed Description

Using formulas, our graph NN architecture of type 1 uses the following propagation rule:

$$L_0 = In \tag{9.1}$$

$$L_{2i+1} = \phi \left(W_{2i+1}^T L_{2i} + C_{2i+1} \right) \tag{9.2}$$

$$L_{2i+2} = \phi \left(W_{2i+2} L_{2i+1} + B_{2i+2} \right) \tag{9.3}$$

$$Out = V^T L_n + D \tag{9.4}$$

where ϕ is an activation function, n is the number of layers, In corresponds to input $N \times 1$ vector, L_{2i} is a vertex-localized $N \times 1$ vector (input to the $2i$-th layer) and L_{2i+1} is an edge-localized $M \times 1$ vector (input to the $2i + 1$-th layer), where M is the number of edges in our problem graph. The core element of the propagation rule are the trainable weight matrices W_j of shape $N \times M$, which fulfill the property

$$\left(W_j \right)_{kl} \neq 0 \Leftrightarrow \text{vertex } k \text{ is one of the two ends of edge } l. \tag{9.5}$$

Additionally, B_j and C_j are trainable bias vectors, and there is a weight and bias pair (V, D) for the output layer. Figure 9.1 shows an exemplary road network and the corresponding matrix W structure (rows correspond to vertices and columns to edges of the crossroad adjacency graph).

Fig. 9.1 An exemplary road network and the corresponding matrix W. Crossroads are denoted by numbers from 1 to 6 and edges by letters from a to f

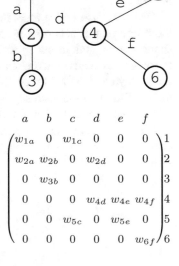

$$
\begin{array}{cccccc}
a & b & c & d & e & f \\
\end{array}
$$

$$
\begin{pmatrix}
w_{1a} & 0 & w_{1c} & 0 & 0 & 0 \\
w_{2a} & w_{2b} & 0 & w_{2d} & 0 & 0 \\
0 & w_{3b} & 0 & 0 & 0 & 0 \\
0 & 0 & 0 & w_{4d} & w_{4e} & w_{4f} \\
0 & 0 & w_{5c} & 0 & w_{5e} & 0 \\
0 & 0 & 0 & 0 & 0 & w_{6f}
\end{pmatrix}
\begin{array}{c}
1 \\ 2 \\ 3 \\ 4 \\ 5 \\ 6
\end{array}
$$

For the architecture of type 2, the propagation rule is even simpler:

$$L_0 = In \tag{9.6}$$

$$L_i = \phi\left(A_i L_{i-1} + B_i\right) \tag{9.7}$$

$$Out = V^T L_n + D \tag{9.8}$$

where all L_i's are now $N \times 1$ vectors and the A_i's are trainable $N \times N$ weight matrices that fulfill the property

$$(A_i)_{kl} \neq 0 \Leftrightarrow \text{vertex } k \text{ is adjacent to vertex } l. \tag{9.9}$$

The meaning of B_i, V and D is the same as in Eqs. (9.1)–(9.4).

One may notice that the definition above covers only cases with a single feature/channel per edge/vertex. In order to extend it to the cases with n_j features, one may use the following propagation rule:

$$L_0 = In \tag{9.10}$$

$$(L_{2i+1})_{rs} = \phi\left(\sum_{t,u} (W_{2i+1})_{tsru} (L_{2i})_{tu} + (C_{2i+1})_{rs}\right) \tag{9.11}$$

$$(L_{2i+2})_{rs} = \phi\left(\sum_{t,u} (W_{2i+2})_{rstu} (L_{2i+1})_{tu} + (B_{2i+2})_{rs}\right) \tag{9.12}$$

$$Out = V^T L_n + D \tag{9.13}$$

where the W_j's are now 4-dimensional trainable weights tensors of shapes $\left(N, n_j, M, n_{j-1}\right)$ where n_j denotes the number of channels in layer j. These satisfy

$$\left(W_j\right)_{krls} \neq 0 \Leftrightarrow \text{vertex } k \text{ is one of the two ends of edge } l \tag{9.14}$$

just as in formula (9.5). Additionally, the biases B_j and C_h now have to be matrices instead of vectors. The intuitive explanation of formulas (9.10)–(9.14) is that the entries of the matrix W_j become $n_j \times n_j$ matrices instead of numbers. The sparse connectivity pattern is kept for the edge/vertex related indices, while there is full connectivity in the feature/channel dimension.

For graph neural nets of type 2, the formulas analogous to (9.10)–(9.13) are:

$$L_0 = In \tag{9.15}$$

$$(L_i)_{rs} = \phi\left(\sum_{t,u} (A_i)_{rstu} (L_{i-1})_{tu} + (B_i)_{rs}\right) \tag{9.16}$$

$$Out = V^T L_n + D \tag{9.17}$$

where the trainable four-dimensional weight tensors satisfy a property analogous to Formula (9.9):

$$(A_i)_{krls} \neq 0 \Leftrightarrow \text{ vertex } k \text{ is adjacent to vertex } l. \tag{9.18}$$

The number of channels that turned out sufficient to obtain a good fit in our case is $3-4$ (to explore the expressive models territory, we also considered 5 channels in our main experiment). Also, $3-4$ layers yielded enough expressive power for our problem. The activation function ϕ we considered was tanh, as in the previous studies [6, 14], tanh worked much better than ReLU in terms of generalization properties to unseen minima of the simulator output. Also, our preliminary experiments with GNN led to the same conclusions.

The tensors W_j and/or A_j were initialized using Glorot initialization [2].

9.3.2 Graph Neural Networks of Type 1—Initial Motivation

The motivation behind the alternating vertex/edge layers for graph neural networks of type 1 may not be obvious at first sight. We initially arrived at this idea by considering a very basic model of how well two neighbouring crossroads are synchronized in terms of producing short average waiting times at a red lights. The rough conclusion of such analysis is that the average waiting time would depend on the quality of the match between the average travel time between the two crossroads and the difference between their green light phase shifts. To understand this better, let us imagine that in our road network, there are only two traffic lights with phase shifts s_1 and s_2. Travelling between them takes time t on average. Let us also assume that cars arrive randomly at traffic signal 1 only and travel towards traffic signal 2. Thus, the average waiting time on signal 1 is something we cannot control. However, the average waiting time on signal 2 is something that can be controlled by adjusting the difference between s_1 and s_2. After a brief reflection, the formula for the average waiting time on the two signals can be obtained as

$$l_1 + l_2 - \max\left(\min\left(l_1, l_2\right) - |s_2 - s_1 - t|, 0\right)$$
$$- \max\left(\min\left(l_1, l_2\right) - 120 + |s_2 - s_1 - t|, 0\right), \tag{9.19}$$

where $l_1 < 120$ and $l_2 < 120$ are the lengths of the red light interval for signals 1 and 2, respectively. For simplicity, $|s_2 - s_1 - t|$ is assumed to belong to the interval $[0, 120)$. For other cases, it is enough to consider $s_2 - s_1 - t$ modulo 120.

As we can see, in our context Formula (9.19) is a simple function of input values on the two ends of a road network graph edge. Formula (9.19) is easily expressed using linear combinations and ReLUs. Thus, our guess was that grouping outputs from 'vertex layers' into pairs to form 'edge layers' should help the learned model to be well tailored to our problem. Even though `tanh` turned out to be a better activation than `ReLU`, the general intuition proved to be correct.

9.4   Setup of Experiments

9.4.1   Dataset

The datasets for experiments were generated using the Traffic Simulation Framework (TSF) software [3] for 3 districts in Warsaw: Centrum, Ochota and Mokotów. The inputs to the simulator were vectors of lengths 11, 21 and 42, respectively. Each position in a vector represented an offset of a traffic signal on a corresponding intersection. The offsets are shifts with respect to a global two minute traffic signal cycle start—times from the beginning of the simulation to the first switch from the red signal state to the green signal state (it was assumed for simplicity that the duration of a green signal phase is always equal to 58 s, while duration of a red signal phase is equal to 62, constituting a 120-s cycle). The offsets were provided as integers, measured in seconds, hence they ranged from 0 to 119 (note the periodicity of these variables). In the process of generating datasets, for each position in the input vector values from the set $\{0, 1, \ldots, 119\}$ were sampled from the uniform distribution independently.

The simulator output in each case was the total waiting time on red signals, summed for all the cars participating in the simulation in a considered area (finding the inputs minimizing this output value was the optimization goal of the considered TSS problem instance).

Each simulation lasted 10 minutes and consisted of 42000 cars on the whole road network of Warsaw. The datasets for the considered districts in Warsaw (Ochota, Mokotów and Centrum) were generated using approximately 100000 randomly selected inputs for the TSF simulator. These datasets are publicly available to enable further research [20].

Output values are roughly in the range from 38000 to 60000 for Ochota, 285000–330000 for Mokotów and 67000–85000 for Centrum. The exact distributions of outputs in the three datasets are shown in Figs. 9.2, 9.3 and 9.4. The minimum simulator outputs in the datasets are: 37838 for Ochota, 285483 for Mokotów and 66742 for Centrum. These values are important because the optimization algorithms applied to the TSS problem (such as gradient descent) had the goal to find traffic signal settings (inputs to the simulator) minimizing the TSF's output.

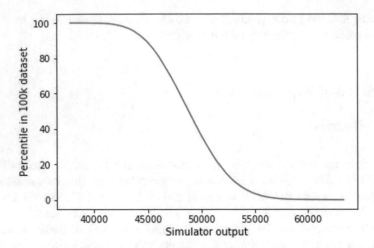

Fig. 9.2 Cumulative distribution function of the TSF's output values for the dataset generated for Ochota district. Volume corresponding to near-optimal parameter settings is extremely low

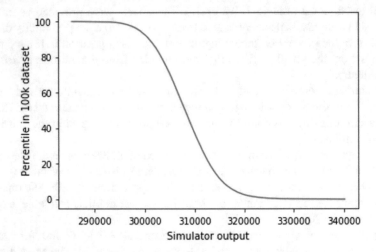

Fig. 9.3 Cumulative distribution function of the TSF's output values for dataset generated for the Mokotów district. Volume corresponding to near-optimal parameter settings is extremely low

9.4.2 Training and Evaluation

After preparing the datasets, we trained GNN and FCNN networks as metamodels to solve TSS using gradient descent.

Before training, we scaled the values at positions in the input vectors to $[-1, 1]$. The values in each input vector were transformed by $x \mapsto \sin(2\pi x/120)$ and $x \mapsto \cos(2\pi x/120)$ mapping, thus doubling the input size (actually, increasing its number of channels to 2 in the case of GNN). This is motivated by the periodicity of the

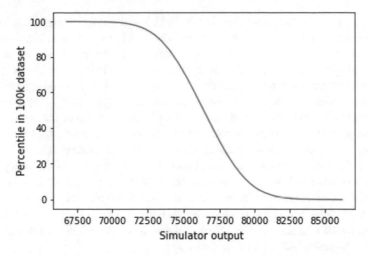

Fig. 9.4 Cumulative distribution function of the TSF's output values for the dataset generated for the Centrum district. Volume corresponding to near-optimal parameter settings is extremely low

problem—it turned out that thanks to such a transformation, the neural network may learn that the offsets are periodic and values 0 and 120 correspond to the same setting [6, 14]. For the output, we used a standard scaler that divides the data by its standard deviation and shifts the mean to zero.

For each of the 3 considered road networks, we tested 9 different GNN architectures (cf. Sects. 9.3 and 9.3.1 for a detailed description of GNN architectures) and 9 FCNN architectures.

The 9 selected GNN architectures corresponded to all the possible combinations of values from the following two parameter sets:

- number of hidden GNN layers: 2, 3, 4 (not counting input and output layers),
- number of channels per layer: 3, 4, 5.

The activation function we used was tanh, indicated as superior to ReLU in preliminary experiments, as well as in previous works on this problem [6, 14].

For comparison, we tested also 9 different FCNN architectures, covering all the possible combinations of values from the following two parameter sets:

- number of hidden layers: 2, 3, 4,
- number of neurons per layer: 20, 40, 100.

The number of neurons per layer was chosen based on some preliminary experimentation. The activation function was tanh again.

For each of the 3 datasets, we used the same 90/10 train/test split for each of the considered 18 hyperparameter settings (9 GNN architectures and 9 FCNN architecture). For each of the architectures, we ran the following procedure:

1. Train a model on the training set for about 1100 epochs (concretely, minimize on 10^5 random mini-batches of size 997 (997 is the closest prime to 1000—a prime number was chosen to assure better randomization, although it was not expected to have any real effect) using Adam optimizer [10] and a learning rate of 0.0035).
2. Evaluate the trained model on the test set using the mean relative error with respect to the original outputs (before normalization and scaling) as the core metrics.
3. Generate 100 gradient descent trajectories (for 3000 steps) of the trained model output with respect to its inputs (in the original input space, backpropagating through the sin/cos transformation). Gradients were evaluated at inputs rounded in the original parameter space (the applied traffic simulator (TSF) accepts only integer values of offsets). Nesterov updater [15] with a learning rate of 0.01 and momentum of 0.9 is used. This is similar to approach used in [14].
4. Every 30 steps, transform the current trajectory point to the original parameter space, round and send to the TSF simulator. Save the inputs and the simulator outputs to a new 'simulation' test set.
5. Evaluate the trained model on the 'simulation' test set using various metrics (cf. the discussion in Sect. 9.5).

Please note that our 'test' set does not play a typical role of a test set in ML experiments. Instead, we use it only to define an order on the considered GNN and FCNN architectures, for use in tables and plots (e.g., Table 9.1). The actual 'testing' is performed using the 'simulation' datasets (cf. point 5 of the procedure described earlier). These datasets are different for each model, as gradient descent trajectories are also different.

All the experiments were run on virtual machines in the Azure cloud (NC6 with NVIDIA Tesla K80 [21]). The code used in the experiments can be found at [19]. All of the models trained for the main experiment and all the out-of-sample simulation datasets (cf. Sect. 9.4 for more details) can be found at [22], while the core dataset can be obtained at [20] to ensure reproducibility of the results and enable further research in this domain.

9.5   Experiment Results

The key results of our experiments with GNN are shown in Tables 9.1, 9.2, 9.3, 9.4 and 9.5, as well as in Figs. 9.5, 9.6 and 9.7.

Table 9.1 shows a summary of core performance measures, calculated for three top GNN and three top FCNN, ranked based on the average accuracy (Mean Absolute Percentage Error—MAPE) on the test set. The core presented measures are:

• Minimum MAPE on the test set. This number can be obtained *before* performing gradient descent. The minimum is taken among the three top-ranked GNN or FCNN (according to the row description). Because of the model selection criterion, we use for Table 9.1, this minimum is global within the respective 9-element model universe (GNN or FCNN).

Table 9.1 Core results for the three best GNN and the three best FCNN architectures according to the accuracy (MAPE) on the test set (i.e., gradient descent results did not affect the selection of these models)

Measure	Model	Ochota	Mokotów	Centrum
Min. MAPE	GNN	1.33%	0.76%	0.80%
On the test set	FCNN	1.71%	0.94%	0.87%
Min. simulation output	GNN	32,205	265,129	63,606
	FCNN	32,587	266,237	63,553
Avg MAPE on the lowest	GNN	1.26%	0.53%	0.76%
5% sim. outputs	FCNN	5.35%	3.04%	2.49%
Avg MAPE on the lowest	GNN	1.75%	0.84%	1.22%
10% sim. outputs	FCNN	4.53%	2.74%	2.25%
Avg MAPE on the lowest	GNN	1.51%	1.00%	1.11%
15% sim. outputs	FCNN	4.65%	2.56%	2.04%

- Minimum simulation output obtained when performing gradient descent (note that while being interesting from the traffic optimization perspective, this measure lacks robustness, as it can be distorted by a single data point).
- Average MAPE on $x\%$ (for $x = 5, 10, 15$) gradient descent trajectory ends, selected according to the corresponding simulator output value (sorted lowest first). To arrive at the values presented in rows 5–10 of Table 9.1, an average is taken over the three models selected, GNN or FCNN, according to the row description.

First, let us note that the results of Table 9.1 show a better performance of GNN comparing to FCNN, particularly in terms of minimum MAPE on the test set and average MAPE on the lowest points from the gradient descent trajectory according to the simulation. The improvement is visible for all the three road networks (Ochota, Mokotów, Centrum) and all the five core measures (with the exception of the minimum simulation output value obtained for Centrum, where one FCNN turned out to yield slightly (less than 0.1%) lower result than all the GNN).

To summarize, the core improvement areas are:

- Much lower error on the test set.
- Lower minimum simulator output value obtained when doing the gradient descent (except for Centrum, for which we can count a draw).
- Much lower approximation error obtained on the trajectory ends corresponding to 5%, 10% and 15% lowest simulator output values.

For performance measure calculations in Table 9.1, we selected the three GNN and the three FCNN that yielded the best test set performance, for each of the three

Table 9.2 Core results for the best GNN and the best FCNN according to test set accuracy

Measure	Model	Ochota	Mokotów	Centrum
Minimal simulation	GNN	32,299	265,129	63,606
Output	FCNN	33,098	266,237	63,941
Avg MAPE on the lowest	GNN	0.51%	0.41%	1.69%
5% simulation outputs	FCNN	1.90%	3.18%	2.07%
Avg MAPE on the lowest	GNN	0.71%	0.66%	1.33%
10% simulation outputs	FCNN	1.75%	3.20%	1.72%
Avg MAPE on the lowest	GNN	0.87%	0.76%	1.29%
15% simulation outputs	FCNN	1.78%	2.93%	1.77%

Table 9.3 Core results for two best GNN and two best FCNN according to test set accuracy

Measure	Model	Ochota	Mokotów	Centrum
Minimum simulation	GNN	32,288	265,129	63,606
Output	FCNN	32,587	266,237	63,553
Avg MAPE on the lowest	GNN	1.85%	1.10%	0.29%
5% simulation outputs	FCNN	2.28%	5.05%	1.52%
Avg MAPE on the lowest	GNN	1.42%	1.02%	0.29%
10% simulation outputs	FCNN	2.15%	4.19%	1.63%
Avg MAPE on the lowest	GNN	1.33%	1.13%	0.36%
15% simulation outputs	FCNN	2.03%	3.88%	1.53%

considered road networks (Centrum, Ochota Mokotów). It is natural to ask if our conclusions are robust against choosing a different number of best performing models. To demonstrate this, we include Tables 9.2, 9.3, 9.4 and 9.5 below. These tables were prepared using one, two, five and nine (a.k.a. all) best-performing GNN and FCNN according to the test set performance. Minimum average test set error is not shown, because, by construction, it would be the same in all these tables, as well as in Table 9.1. It is easily seen that the above conclusions from Sect. 9.5 are upheld.

Table 9.4 Core results for five best GNN and five best FCNN according to test set accuracy

Measure	Model	Ochota	Mokotów	Centrum
Minimum simulation	GNN	32,205	265,129	63,606
Output	FCNN	32,587	266,237	63,553
Avg MAPE on the lowest	GNN	0.99%	0.35%	0.58%
5% simulation outputs	FCNN	8.59%	1.89%	1.92%
Avg MAPE on the lowest	GNN	0.98%	0.44%	0.88%
10% simulation outputs	FCNN	8.59%	1.30%	1.92%
Avg MAPE on the lowest	GNN	0.92%	0.37%	1.01%
15% simulation outputs	FCNN	8.54%	1.01%	1.92%

Table 9.5 Core results for all GNN and all FCNN we considered

Measure	Model	Ochota	Mokotów	Centrum
Minimum simulation	GNN	32,205	265,129	63,606
Output	FCNN	32,587	266,237	63,553
Avg MAPE on the lowest	GNN	0.87%	2.27%	1.39%
5% simulation outputs	FCNN	4.63%	1.12%	1.60%
Avg MAPE on the lowest	GNN	0.87%	2.71%	0.88%
10% simulation outputs	FCNN	4.44%	1.73%	1.79%
Avg MAPE on the lowest	GNN	0.84%	3.03%	0.88%
15% simulation outputs	FCNN	4.30%	1.87%	1.39%

Figures 9.5, 9.6 and 9.7 show the density of gradient descent trajectory points as heatmap plots for 3 best GNN/FCNN models for each district. The horizontal axes correspond to gradient descent trajectory point numbers (each trajectory had 3000 steps, but we recorded points every 30 steps), and the vertical axes correspond to respective simulator output values. The plots simply show a heatmap plot of these points on the (point number, simulator output) plane. Thus, the more points in some

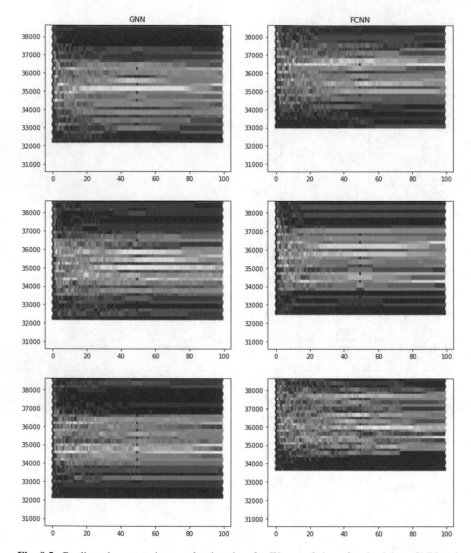

Fig. 9.5 Gradient descent trajectory density plots for Warsaw Ochota for the 3 best GNN and FCNN models (charts on the left correspond to GNN, charts on the right to FCNN). Horizontal axis corresponds to trajectory point number (recorded every 30 steps), vertical axis to simulator output value. The more points in some areas, the brighter the colour

area, the brighter the color. Also, if one architecture happened to reach a lower minimum than another, the resulting heatmap is taller.

Besides confirming some of the quantitative conclusions from Tables 9.1, 9.2, 9.3, 9.4 and 9.5, the heatmap plots also show that in many cases, the gradient descent is less 'noisy' for GNN, suggesting a smoother function surface, less prone to overfit noise.

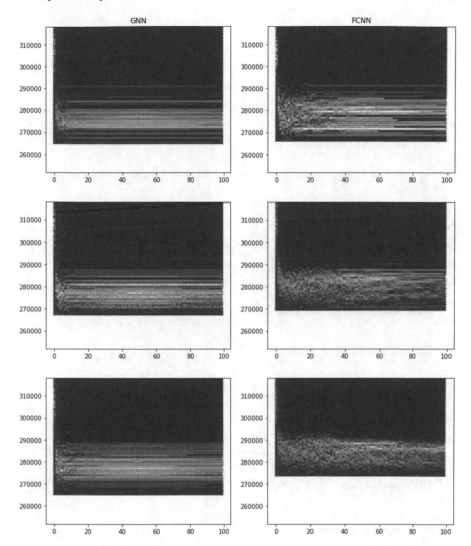

Fig. 9.6 Gradient descent trajectory density plots for Warsaw Mokotów for the 3 best GNN and FCNN models (charts on the left correspond to GNN, charts on the right to FCNN). Horizontal axis corresponds to trajectory point number (recorded every 30 steps), vertical axis to simulator output value. The more points in some areas, the brighter the colour

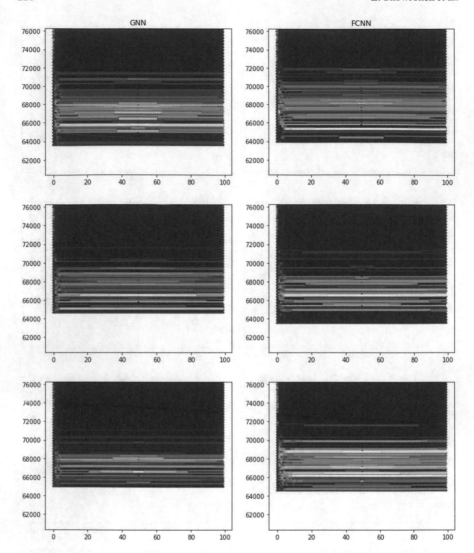

Fig. 9.7 Gradient descent trajectory density plots for Warsaw Centrum for the 3 best GNN and FCNN models (charts on the left correspond to GNN, charts on the right to FCNN). Horizontal axis corresponds to trajectory point number (recorded every 30 steps), vertical axis to simulator output value. The more points in some areas, the brighter the colours

Based on the evidence presented, we concluded that the introduced graph-based neural networks:

- Generalize to the original fixed test sets better than FCNN.
- Generalize significantly better than FCNN in the closeness of the optimization target function minima.
- Allow finding inputs corresponding to small simulator output values more easily. Most of the time, the values obtained are also lower than those obtained with a help of FCNN.
- Qualitatively produces smoother gradient descent trajectories that are less prone to overfitting, for multiple hyperparameter settings.

It is also worth mentioning that the simulator output values obtained when doing the gradient descent runs are significantly lower than ones that could likely be obtained by random search algorithms. In our training datasets (of sizes approximately equal 100k), the minimum simulator output values were around 38000, 285000 and 67000 for Ochota, Mokotów and Centrum, respectively (with the mean values of approximately 49000, 308000, 76000, c.f., Figs. 9.2, 9.3 and 9.4) whereas the gradient descent method steadily achieved (based on 5%, 10% and 15% percentiles), results below 33000, 270000 and 65000, respectively (with acceptable average relative errors: 1.65%, 0.64% and 1.68%, respectively).

9.6   Consistency Checks

The findings of the previous section call for some careful consistency checks before reaching final conclusions. In particular, it is not fully clear that the actual adjacency matrix brings any value. Perhaps, using any similar sparse graph, even not related to the problem at hand, could result in similar performance.

Therefore, we decided to run some additional experiments where the original graph topology was perturbed (either by turning some edges to non-edges and vice versa, or by permuting graph vertex labels). The other steps of the model fitting procedure were kept exactly the same as before. The results we obtained indicate that the mean squared error achieved on (fixed) test set grows approximately monotonously as a function of the distance (symmetric difference between edge sets) of the graph we use (to build the neural network) to the actual problem graph. This implies that the information about the actual graph topology is important for achieving better accuracy in the main set of experiments (cf. Figs. 9.8, 9.9, 9.10, 9.11, 9.12 and 9.13).

In detail, the procedure we used was the following. First, we decided to fix the number of layers to 3 and the number of channels to 4 per layer (for GNN of type 1). Then, we built our nets using random graphs with various degrees of resemblance to the original problem graph. We repeated this for all the three road networks considered in this paper. As a measure of graph similarity, we used the symmetric difference between the sets of graph edges. The random graphs were generated in

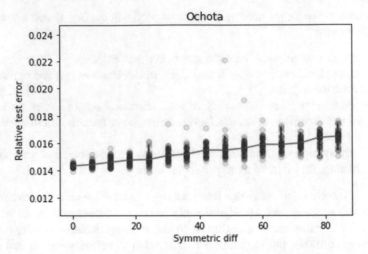

Fig. 9.8 Mean relative error achieved by a GNN on the test set for Ochota after roughly 330 epochs of training, plotted as a function of the distance of a random graph to the true one. Edge/non-edge switching (described in the text) was used to generate random graphs. Errorbars correspond to 5% sample quantiles. Red dots denote the median result

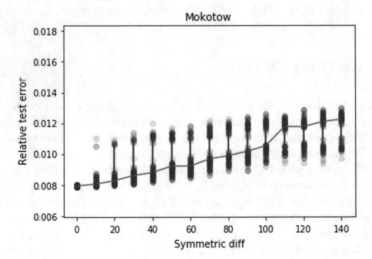

Fig. 9.9 Mean relative error achieved by a GNN on the test set for Mokotów after roughly 330 epochs of training, plotted as a function of the distance of a random graph to the true one. Edge/non-edge switching (described in the text) was used to generate random graphs. Errorbars correspond to 5% sample quantiles. Red dots denote the median result

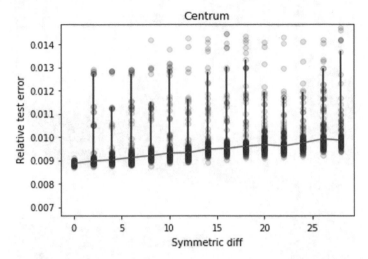

Fig. 9.10 Mean relative error achieved by a GNN on the test set for Centrum after roughly 330 epochs of training, plotted as a function of the distance of a random graph to the true one. Edge/non-edge switching (described in the text) was used to generate random graphs. Errorbars correspond to 5% sample quantiles. Red dots denote the median result

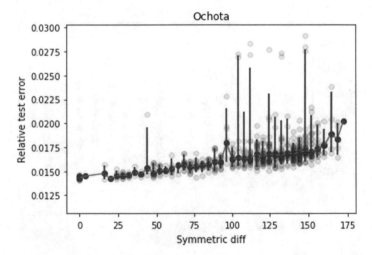

Fig. 9.11 Mean relative error achieved by a GNN on the test set for Ochota after roughly 330 epochs of training, plotted as a function of the distance of a random graph to the true one. Vertex label permutation (described in the text) was used to generate random graphs. Errorbars correspond to 5% sample quantiles. Red dots denote the median result

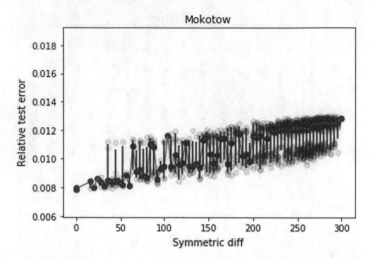

Fig. 9.12 Mean relative error achieved by a GNN on the test set for Mokotów after roughly 330 epochs of training, plotted as a function of the distance of a random graph to the true one. Vertex label permutation (described in the text) was used to generate random graphs. Errorbars correspond to 5% sample quantiles. Red dots denote the median result

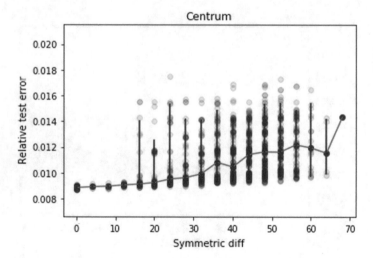

Fig. 9.13 Mean relative error achieved by a GNN on the test set for Centrum after roughly 330 epochs of training, plotted as a function of the distance of a random graph to the true one. Vertex label permutation (described in the text) was used to generate random graphs. Errorbars correspond to 5% sample quantiles. Red dots denote the median result

two ways. The first method (referred to later as **'Edge/Non-edge switching'**) used random edge insertions and deletions, with the desired value of the symmetric difference kept fixed. The second method (referred to later as **'Vertex label permutation'**) used random permutations of the vertex labels while keeping the connection graph structure exactly the same. Graphs generated by this method were isomorphic, but not identical to the original one.

It is worth mentioning that although the first method generates truly random graphs similar to the original one, the new graph might not represent a plausible road network. The second method, on the contrary, always keeps the same, realistic road network graph structure, but it provides spurious insights to the training algorithm as crossroads are switched.

The results obtained by these two methods are shown in Figs. 9.8, 9.9 and 9.10, corresponding to Ochota, Mokotów and Centrum datasets, respectively. The plots show the mean relative error achieved on the test sets by neural nets based on random graphs, after roughly 330 epochs of training. The values on the horizontal axis correspond to the distance of the graph used for constructing the net to the true graph. The distance was measured using the symmetric difference between the two graphs' edge sets. For each training run, the same train/test split of the respective dataset was used, with a train/test proportion of 90/10. However, we allowed the weights to be fully randomly initialized. It should be noted that in all the cases we had manually checked, the training and test errors after $100-200$ epochs had already been stable, so the number of $\sim$330 epochs was chosen with some security margin. As we can see, the median of the mean relative error, denoted with a red dot, grows almost monotonously as a function of the distance of the graph we use to the actual problem graph. This is visible for both graph sampling methods. The minimum average relative error attained for a particular value of the distance also grows, perhaps with a bit more of noise.

9.7 Conclusions and Future Work

We demonstrated the usefulness of sparsely connected neural networks, with sparsity based on an adjacency graph, in a problem from the traffic optimization domain. GNN consistently outperformed FCNN on fixed test sets for the three realistic road networks we considered (Ochota, Mokotów and Centrum districts in Warsaw). More importantly, GNN achieved approximation quality far superior to FCNN near unseen simulator output value minima. By using randomly perturbed graphs, we also showed that the choice of the proper graph when constructing a GNN is important for achieving good results on a test set.

The kind of NN sparsity considered in this paper, where only some of the connections are allowed, may be regarded as a kind of regularizer based on the problem graph. It is similar to L1 regularization of a fully connected neural network in that it keeps only some weights non-zero in the trained model. The resulting networks have much fewer parameters than analogous fully connected networks and turn out to generalize significantly better than any architecture we considered so far for solving the TSS problem.

As for future work, we would like to investigate how to ensure good scalability and parallelizability of GNN for a purpose of traffic optimization on larger scales. Another interesting research direction is to use transfer learning to train GNN models to approximate simulations for different traffic conditions on the same road networks. Thanks to the graph-based neural network architecture, it may be possible to investigate the explainability properties of these models, e.g., to see which intersections have the greatest impact on road traffic conditions on a given area or on another intersection.

Acknowledgements The presented research was supported by Microsoft's "AI for Earth" computational grant.

References

1. Chenyi, Z., Qiang, M.: Dual graph convolutional networks for graph-based semi-supervised classification. In: Proceedings of the 2018 World Wide Web Conference, pp. 499–508 (2018)
2. Glorot, X., Bengio, Y.: Understanding the difficulty of training deep feedforward neural networks. In: Proceedings of the Thirteenth International Conference on Artificial Intelligence and Statistics. PMLR, vol. 9, pp. 249–256 (2010)
3. Gora, P.: Traffic simulation framework - a cellular automaton based tool for simulating and investigating real city traffic. In: Recent Advances in Intelligent Information Systems, pp. 641–653 (2009)
4. Gora, P., Bardoński, M.: Training neural networks to approximate traffic simulation outcomes. In: 2017 5th IEEE International Conference on Models and Technologies for Intelligent Transportation Systems (MT-ITS), pp. 889–894. IEEE (2017)
5. Gora, P., Kurach, K.: Approximating traffic simulation using neural networks and its application in traffic optimization (2016). arXiv:abs/2102.12896
6. Gora, P., Brzeski, M., Możejko M., Klemenko, A., Kochański, A.: Investigating performance of neural networks and gradient boosting models approximating microscopic traffic simulations in traffic optimization tasks (2018). arXiv:abs/1812.00401
7. Guo, S., Lin, Y., Feng, N., Song, C., Wan, H.: Attention based spatial-temporal graph convolutional networks for traffic flow forecasting. In: Proceedings of the AAAI Conference on Artificial Intelligence, vol. 33(01), pp. 922–929 (2019)
8. Henderson, S.G., Nelson, B.L.: Chapter 18 metamodel-based simulation optimization. In: Handbooks in Operations Research and Management Science, Elsevier, Russell R. Barton and Martin Meckesheimer, vol. 13, pp. 535–574 (2006)
9. Hurrion, R.D.: A sequential method for the development of visual interactive meta-simulation models using neural networks. J. Oper. Res. Soc. **6**(51), 712–719 (2000)
10. Kingma, D.P., Ba, J.: Adam: a method for stochastic optimization. In: Proceedings of the 3rd International Conference on Learning Representations (ICLR) (2015)
11. Kipf, T.N., Welling, M.: Semi-supervised classification with graph convolutional networks. In: Proceedings of the 5th International Conference on Learning Representations (2017)
12. Li, Y., Yu, R., Shahabi, C., Liu, Y.: Graph convolutional recurrent neural network: data-driven traffic forecasting. CoRR (2017). arXIv:abs/1707.01926
13. Micheli, A.: Neural network for graphs: a contextual constructive approach. IEEE Trans. Neural Netw. **20**(3), 498–511 (2009)

14. Możejko, M., Brzeski, M., Mądry, Ł., Skowronek, Ł., Gora, P.: Traffic signal settings optimization using gradient descent. Schedae Informaticae **27** (2018)
15. Nesterov, Y.: A method for unconstrained convex minimization problem with the rate of convergence o $(1/k^2)$. Doklady AN USSR. **269**, 543–547 (1983)
16. OpenStreetMap Service. https://www.openstreetmap.org, Cited 22 July 2022
17. Osorio, C., Bierlaire, M.: A surrogate model for traffic optimization of congested networks: an analytic queueing network approach. Technical report TRANSP-OR090825. Transport and Mobility Laboratory, ENAC, EPFL (2009)
18. Rijnen, D., Rhuggenaath, J., Costa, P.R.D.O.D., Zhang, Y.: Machine learning based simulation optimisation for trailer management. In: 2019 IEEE International Conference on Systems, Man and Cybernetics (SMC), pp. 3687–3692 (2019)
19. Skowronek, Ł., Gora, P., Możejko, M., Klemenko, A.: Zipped repository of the code used in experiments. https://drive.google.com/file/d/1FF6q8GTJljkYjKSNMYsL5neoXbOqPIcm/view?usp=sharing, Cited 22 July 2022
20. Skowronek, Ł., Gora, P., Możejko, M., Klemenko, A.: Dataset used in experiments. https://drive.google.com/file/d/1aLUL3QPxGxeUVmqds6HWeGnVnQ5O4Mxr/view?usp=sharing, Cited 22 July 2022
21. Skowronek, Ł., Gora, P., Możejko, M., Klemenko, A.: Description of virtual machines used in experiments. https://docs.microsoft.com/en-us/azure/virtual-machines/sizes-gpu, Cited 22 July 2022
22. Skowronek, Ł., Gora, P., Możejko, M., Klemenko, A.: Models trained for the main experiment and all the out-of-sample simulation datasets. https://drive.google.com/file/d/1mPnFt1Y1wGLGE-ha2_JiYsuJEebnfBYx/view?usp=sharing, Cited 22 July 2022
23. Skowronek, Ł., Gora, P., Możejko, M., Klemenko, A.: Toy dataset used in experiments. https://drive.google.com/file/d/1y-25uIiPRQb7zUNQshrh8UXtegLKg25e/view?usp=sharing, Cited 22 July 2022
24. Yang, T.B., Yeh, Y.J.: The model and properties of the traffic light problem. In: Proceedings of International Conference on Algorithms, pp. 19–26 (1996)
25. Yao, H., Wu, F., Ke, J., Tang, X., Jia, Y., Lu, S., Gong, P., Ye, J., Li, Z.: Deep multi-view spatial-temporal network for taxi demand prediction. CoRR (2018). arXiv:abs/1802.08714
26. Yu, B., Yin, H., Zhu, Z.: Spatio-temporal graph convolutional neural network: a deep learning framework for traffic forecasting. CoRR (2017). arXiv:abs/1709.04875
27. van der Maaten, L., Hinton, G.: Visualizing data using t-SNE. J. Mach. Learn. Res. 2579–2605 (2008)
28. Wu, Z., Pan, S., Chen, F., Long, G., Zhang, C., Yu, P.S.: A comprehensive survey on graph neural networks. IEEE Trans. Neural Netw. Learn. Syst. **32**(1), 4–24 (2019)
29. Zhang, J., Chowdhury, S., Zhang, J., Messac, A., Castillo, L.: Adaptive hybrid surrogate modeling for complex systems. AIAA J. 643–656 (2013)

Index